SpringerBriefs in Electrical and Computer Engineering

SpringerBriefs present concise summaries of cutting-edge research and practical applications across a wide spectrum of fields. Featuring compact volumes of 50 to 125 pages, the series covers a range of content from professional to academic. Typical topics might include: timely report of state-of-the art analytical techniques, a bridge between new research results, as published in journal articles, and a contextual literature review, a snapshot of a hot or emerging topic, an in-depth case study or clinical example and a presentation of core concepts that students must understand in order to make independent contributions.

More information about this series at http://www.springer.com/series/10059

Yue Gao • Zhijin Qin

Data-Driven Wireless Networks

A Compressive Spectrum Approach

 Springer

Yue Gao
School of Electronic Engineering
and Computer Science
Queen Mary University of London
London, UK

Zhijin Qin
School of Electronic Engineering
and Computer Science
Queen Mary University of London
London, UK

ISSN 2191-8112   ISSN 2191-8120   (electronic)
SpringerBriefs in Electrical and Computer Engineering
ISBN 978-3-030-00289-3   ISBN 978-3-030-00290-9   (eBook)
https://doi.org/10.1007/978-3-030-00290-9

Library of Congress Control Number: 2018956141

This Springer imprint is published by the registered company Springer Nature Switzerland AG
The registered company address is: Gewerbestrasse 11, 6330 Cham, Switzerland

How can wireless data influence the compressive spectrum sensing in windband cognitive radio and Internet of Things networks?

Foreword

It is my great pleasure to write a foreword to this fantastic book, *Data-Driven Wireless Networks: A Compressive Sensing Approach*, authored by two of my good friends and long-term collaborators. I first knew the real meaning of "wideband compressive sensing" several years ago after I discussed with them on research in cognitive radio. Later on, I read some of their papers in the related topics and gradually realized that it is a very effective tool in signal processing and wireless communications. I once thought that it would be great to have a comprehensive book in the area.

This book addresses spare representation in wireless communications, with emphasis on the most recently developed compressive sensing-enabled approaches. It starts from a comprehensive overview of the fundamental principles. Subsequently, it introduces the data-driven compressive spectrum sensing in cognitive radio networks and discusses robust and security issues. The book also contains test results of various algorithms based on the real-world signals and data collected by experiments carried out during the TV white space pilot trials.

I would like to emphasize that the authors are very active in the related areas in the past several years. Their publications have been very well-cited and received best paper awards. I believe that this book is as good as their other publications. This book contains most of the important results in the related topics, including their own. The book is written concisely and clearly. It is an excellent reference for graduate students, faculty, and practice engineers in the area.

Professor and Director   Geoffrey Ye Li
Information Transmission and Processing Lab
School of Electrical and Computer Engineering
Georgia Institute of Technology
Atlanta, GA, USA
June 2018

Preface

In this book, we will discuss the applications of spare representation in wireless communications, with particular focus on the most recently developed compressive sensing-enabled approaches. With the help of sparsity property, sub-Nyquist sampling can be achieved in wideband cognitive radio networks by adopting compressive sensing. This book starts from a comprehensive overview of CS principles. Subsequently, we will present a complete framework for data-driven compressive spectrum sensing in cognitive radio networks, which is able to provide guarantee on robustness, low complexity, and security. Particularly, robust compressive spectrum sensing, low-complexity compressive spectrum sensing, and secure compressive sensing-based malicious user detection are proposed to address the various issues in wideband cognitive radio networks. Correspondingly, the real-world signals and data collected by experiments carried out during TV white space pilot trial enable data-driven compressive spectrum sensing. The collected data are used to verify our designs and provide significant insights on the potential of applying compressive sensing to wideband spectrum sensing. We believe this book will provide readers a clear picture on how to exploit the compressive sensing to process wireless signals in wideband cognitive radio networks.

London, UK   Yue Gao
June 2018

Acknowledgment

I would like to express my sincere gratitude to my current and former PhD students, my collaborators, and all the colleagues who contributed to the work and projects that lead to this book. I would like to thank the UK Engineering and Physical Sciences Research Council (EPSRC) for funding my research projects, e.g., the EPSRC Fellowship project (EP/R00711X/1).

I would also like to particularly thank our editor as well as all the editorial staff from Springer in producing this book.

Finally, I would like to thank my family, including my wife, my daughter, and my son. I cannot make it without their great support.

Contents

Part I   Background

1   Introduction 3
   1.1   Motivations and Contributions 4
      1.1.1   Data-Driven Compressive Spectrum Sensing 5
      1.1.2   Robust Compressive Spectrum Sensing 5
      1.1.3   Secure Compressive Spectrum Sensing 6
   References 7

2   Sparse Representation in Wireless Networks 9
   2.1   Principles of Standard Compressive Sensing 9
      2.1.1   Sparse Representation 10
      2.1.2   Projection 10
      2.1.3   Signal Reconstruction 12
   2.2   Reweighted Compressive Sensing 13
   2.3   Distributed Compressive Sensing 14
   2.4   Compressive Spectrum Sensing 15
      2.4.1   Spectrum Sensing Methods 15
      2.4.2   Spectrum Sensing Model 16
      2.4.3   Compressive Wideband Spectrum Sensing 17
   2.5   Summary 19
   References 19

Part II   Compressive Spectrum Sensing Algorithms

3   Data-Driven Compressive Spectrum Sensing 23
   3.1   Introduction 23
      3.1.1   Related Work 24
      3.1.2   Contributions 25
   3.2   Data-Driven Compressive Spectrum Sensing Framework 25
      3.2.1   Iteratively Reweighted Least Square-Based
            Compressive Sensing 26

        3.2.2   Non-iteratively Reweighted Least Square-Based
                Compressive Sensing 28
        3.2.3   Proposed Wilkinson's Method-Based DTT Location
                Probability Calculation Algorithm 31
    3.3 Numerical Analyses... 33
        3.3.1   Numerical Analyses on Simulated Signals and Data 33
        3.3.2   Numerical Analyses on Real-World Signals and Data........ 38
    3.4 Summary 39
    References 40

4   Robust Compressive Spectrum Sensing 43
    4.1 Introduction 43
        4.1.1   Related Work.. 43
        4.1.2   Contributions.. 44
    4.2 Robust Compressive Spectrum Sensing at Single User 45
        4.2.1   System Model... 45
        4.2.2   Computational Complexity and Spectrum Usage Analyses ... 47
    4.3 Numerical Analyses for Single User Case 49
        4.3.1   Analyses on Simulated Signals 49
        4.3.2   Analyses on Real-World Signals............................... 52
    4.4 Matrix Completion-Based Robust Spectrum Sensing at
         Cooperative Multiple Users 53
        4.4.1   System Model... 54
        4.4.2   Denoised Cooperative Spectrum Sensing Algorithm 57
        4.4.3   Computational Complexity and Performance Analyses 58
    4.5 Numerical Analyses for Cooperative Multiple Users Case 58
        4.5.1   Analyses on Simulated Signals 58
        4.5.2   Analyses on Real-World Signals............................... 61
    4.6 Summary 62
    References 63

5   Secure Compressive Spectrum Sensing..................................... 65
    5.1 Introduction 65
        5.1.1   Related Work.. 66
        5.1.2   Motivations and Contributions 67
    5.2 System Model 68
        5.2.1   Networks Description 68
        5.2.2   Signal Processing Model 70
    5.3 Malicious User Detection Framework 71
        5.3.1   Proposed Malicious User Detection Algorithm 72
        5.3.2   Rank Order Estimation Algorithm 75
        5.3.3   Malicious User Number Estimation 78
        5.3.4   Analyses on Minimal Number of Active Secondary Users.... 79
    5.4 Numerical Analyses... 80
        5.4.1   Numerical Results Using Simulated Signals 81
        5.4.2   Numerical Results Using Real-World Signals................. 85

5.5   Summary 86
References 87

Part III   Conclusions

6   Conclusions and Future Work.. 91
6.1   Conclusions 91
6.2   Future Work 92
References 93

Acronyms and Nomenclature

ADC	Analog-to-digital conversion
AOP	Adaptive outlier pursuit
AWGN	Additive white Gaussian noise
CR	Cognitive radio
CRN	Cognitive radio network
CS	Compressive sensing/compressed sensing
CSS	Cooperative spectrum sensing
DNRLS	Data-assisted non-iteratively reweighted least squares
DSA	Dynamic spectrum access
DSO	Digital switch-over
DTT	Digital terrestrial television
DVB-T	Digital Video Broadcasting-Terrestrial
FC	Fusion center
FCC	Federal Communications Commission
FFT	Fast Fourier transform
IDFT	Inverse digital Fourier transform
i.i.d.	Independent and identically distributed
IRLS	Iteratively reweighted least squares
MC	Matrix completion
NGR	National grid reference
NOMA	Non-orthogonal multiple access
Ofcom	Office of Communications
OFDM	Orthogonal frequency division multiplexed
OMP	Orthogonal matching pursuit
PMSE	Programme making and special events
PU	Primary user
QMUL	Queen Mary University of London
RF	Radio frequency
RIP	Restricted isometry property
ROC	Receiver operating characteristics
RTRMC	Riemannian Trust-Region for MC

SNR	Signal-to-noise ratio
SU	Secondary user
TVWS	TV white space
UHF	Ultra-high frequency
WSD	White space device
$\mathbf{d}$	Decision on spectrum occupancy
$\mathscr{F}^{-1}$	Inverse discrete Fourier transform
$h(t)$	Channel coefficients in time domain
$\mathbf{h_f}$	Channel coefficients in frequency domain
$\mathbf{H_f}$	Channel coefficients in frequency domain in matrix format
$\mathscr{I}$	Number of channels among the spectrum of interest
$I_{\max}$	Maximal iteration number
$\mathbf{I}_N$	$N \times N$ identity matrix
J	Number of SUs sensing the same channel at different locations
K	Number of occupied channels, or sparsity level, or rank order of a matrix
$K_{\max}$	Statistical upper bound of the rank order K
$\hat{K}$	Estimated sparsity level K
L	Number of channels in a channel group
L_c	Number of corrupted channels
$\hat{L}_c$	Estimated number of corrupted channels
$\mathbf{M}$	Low-rank matrix
$\hat{\mathbf{M}}$	Recovered low-rank matrix
N	Number of samples at Nyquist rates
p	Norm
p_{ij}	Uncorrupted power value of the ith channel sensed by the jth SU
$\tilde{p}_{ij}$	Corrupted power value of the ith channel sensed by the jth SU
$\hat{p}_{ij}$	Recovered power value of the ith channel sensed by the jth SU
P	Number of compressed measurements
P_{as}	Average received power of the wanted DTT signal
P_d	Probability of detection
$\bar{P}_d$	Target probability of detection
P_f	Probability of false alarm
$\bar{P}_f$	Target probability of false alarm
P_{IB}	Maximum allowable EIRP
$\mathbf{P}^{\Omega}$	Complete matrix constructed at the FC
$r(t)$	Received signal in time domain
$\mathbf{r_f}$	Received signal in frequency domain
$\mathbf{R_f}$	Received signal in frequency domain in matrix format
$s(t)$	Transmitted signal in time domain
$\mathbf{s_f}$	Transmitted signal in frequency domain
$\hat{\mathbf{s}}_\mathbf{f}$	Recovered signal
$\mathbf{S_f}$	Transmitted signal in frequency domain in matrix format
$\hat{\mathbf{S}}_\mathbf{f}$	Recovered signal in matrix format
$w(t)$	Additive white Gaussian noise in time domain
$\mathbf{w_f}$	AWGN in frequency domain

$\mathbf{W_f}$	AWGN in frequency domain in matrix format
$\mathbf{W}$	Weights matrix for IRLS
$\tilde{\mathbf{W}}$	Newly constructed weights matrix for DNRLS
$\mathbf{x}$	Compressed measurements
$\mathbf{X}$	Compressed measurements in matrix format
γ	Compression ratio
κ	Malicious user ratio
λ	Threshold for energy detection
$\mathbf{\Lambda}$	Binary matrix denoting the uncorrupted channels
$\mathbf{\Omega}$	Index set of the complete matrix
$\mathbf{\Phi}$	Measurement matrix
$\mathbf{\Psi}$	Sparsifying matrix
ε	Error tolerance

Part I
Background

Chapter 1
Introduction

Radio frequency (RF) spectrum is a valuable but tightly regulated resource due to its unique and important role in wireless communications. The demand for RF spectrum is increasing due to a rapidly expanding market of multimedia wireless services, while the usable spectrum is becoming scarce due to current rigid spectrum allocation policies. Specifically, according to reports from the Federal Communications Commission (FCC) and the Office of Communications (Ofcom), localized temporal and geographic spectrum utilization is extremely low in reality (Kolodzy and Avoidance 2002; Ofc a). Cognitive radio (CR) has become a promising solution to solve the spectrum scarcity problem, by allowing secondary users (SUs) to opportunistically access a licensed band when the primary user (PU) is absent (Mitola and Maguire 1999). Additionally, it is demonstrated that TV spectrum, which is used to be allocated to analog TV signals, has been cleaned and opened to access due to the digital switch-over (DSO) around the world (Kolodzy and Avoidance 2002; Ofc a). These underutilized TV spectra are named as TV white space (TVWS). Lately, FCC issued a report and order for permitting the cognitive usage of TVWS spectrum (fcc). Most recently, Ofcom has enabled license exempt use of TVWS to harness the benefits of such an innovative wireless technology (Ofc b), which motivates the further research on the cognitive access to TVWS spectrum.

In order to avoid any harmful interference to the PUs in TVWS, SUs in CR networks (CRNs) should be aware of the spectrum occupancy over TV band. Spectrum sensing is the first and one of the most challenging tasks in CR, which is performed to detect the spectrum holes over TV spectrum. As the radio environment changes over time and space, an efficient spectrum sensing technique should be capable of tracking these fast changes (Akyildiz et al. 2006). A good approach for detecting the primary transmitters is to adopt the traditional narrowband sensing algorithms, which include energy detection, matched-filtering, and cyclostationary feature detection. Here, the term "narrowband" implies that the frequency range is sufficiently narrow, such that the channel frequency response can be considered as

© The Author(s), under exclusive license to Springer Nature Switzerland AG 2019
Y. Gao, Z. Qin, *Data-Driven Wireless Networks*, SpringerBriefs in Electrical and Computer Engineering, https://doi.org/10.1007/978-3-030-00290-9_1

flat. In another word, the bandwidth of interest is less than the coherence bandwidth of the channel (Sun et al. 2013).

While the present spectrum sensing algorithms have focused on exploiting spectral opportunities over narrow frequency range, CRNs will eventually be required to exploit spectral opportunities over wide frequency range from hundreds of megahertz (MHz) to several gigahertz (GHz), in order to improve spectrum efficiency and achieve higher opportunistic throughput. In wideband spectrum sensing, as driven by the Nyquist sampling theory, a simple approach is to acquire the wideband signal directly by a high-speed analog-to-digital converter (ADC). So far, wideband spectrum sensing has been investigated in Tian and Giannakis (2007), Farhang-Boroujeny (2008), Quan et al. (2009), Sun et al. (2010) with the implementation of a high-speed ADC. However, the high-speed ADC is particularly challenging or even unaffordable for energy-constrained devices, such as smart phones or even battery-free devices in a wireless power transfer model (Qin et al. 2017). Subsequently, Landau (1967) demonstrated that sampling rate should be no less than the measure of occupied part of the spectrum, with the purpose of guaranteeing the stable reconstruction of multiband signals. However, the energy consumption is still unaffordable for energy-constrained SUs in CRNs. Therefore, revolutionary wideband spectrum sensing techniques become more than desired to release the burden on high-speed ADCs.

Recent developments on compressive sensing (CS) theory inspire sub-Nyquist sampling, by utilizing the sparse nature of signals (Candes 2006). Driven by the inborn nature of the signal sparsity in wireless communications, e.g., the sparse utilization of spectrum, CS theory is capable of enabling sub-Nyquist sampling possible for wideband spectrum sensing. More particularly, CS theory has been firstly applied to wideband spectrum sensing by Tian and Giannakis (2007), where fewer compressed measurements are required on the basis of Nyquist sampling theory. Subsequently, the application of CS theory on wideband spectrum sensing in CRNs has attracted much attention.

1.1 Motivations and Contributions

Along with the developments on CS theory, this book spans the sub-Nyquist-based wideband spectrum sensing with particular emphasis on CS technique. These proposed algorithms are capable of improving the robustness and security of CRNs, with low computational complexity at energy-constrained SUs. The specific motivations and contributions of this book research are summarized in the following.

1.1.1 Data-Driven Compressive Spectrum Sensing

Besides the robustness to channel noise, adaptive compressive spectrum sensing with low complexity has attracted much attention (Zhang et al. 2018). Theoretically, the required number of measurements will proportionally change when the sparsity level of wideband signal varies. However, in practice, the sparsity level of wideband signal is uncertainty, because of either the dynamic activities of PUs or the time-varying fading channels between PUs and SUs. Consequently, most of sub-Nyquist wideband sensing systems should pessimistically choose the number of measurements to ensure exact recovery, leading to more energy consumption at SUs. Moreover, the computational complexity of signal recovery may be unaffordable for the energy-constrained SUs as it is dependent on the number of collected compressed measurements. Therefore, a low-complexity compressive spectrum sensing algorithm is needed, which should be adaptive to the dynamic spectrum occupancy.

Inspired by the geolocation database for TVWS, which is another approach to make SUs aware of spectrum occupancy, a hybrid framework combining compressive spectrum sensing and geolocation database is proposed to achieve adaptive CS with low complexity (Qin et al. 2015, 2016a; Gao et al. 2016). More specifically, a geolocation database algorithm is proposed to be implemented at SUs locally to provide prior information on the spectrum occupancy. As a result, SUs collect samples at the minimum rate without loss of any information. Additionally, with the availability of prior information, a data-assisted non-iteratively reweighted least squares (DNRLS)-based compressive spectrum sensing algorithm is proposed to reduce the computational complexity of signal recovery. In order to further improve accuracy and efficiency of the geolocation database algorithm implemented at SUs, an efficient approach for calculating the maximum allowable equivalent isotropic radiated power (EIRP) is proposed. Furthermore, the proposed hybrid framework and algorithms are tested on the real-world signal and data over TVWS after being approved by the simulated data.

1.1.2 Robust Compressive Spectrum Sensing

With the use of CS at SUs, each SU would only collect compressed samples at sub-Nyquist sampling rate. Subsequently, signal recovery would be performed at SUs or a fusion center (FC), where the data from the spatially located SUs are fused. It is noticed that the signal-to-noise ratio (SNR) of the CS measurements would be decreased by 3 dB for every octave increasing in the subsampling factor for acquisition of a noisy signal with fixed sparsity level (Treichler et al. 2009). This makes the exact signal recovery more difficult for compressive spectrum sensing under heavy channel noise. Therefore, a robust spectrum sensing algorithm based on CS with low computational complexity is needed.

As motivated by this, two robust compressive spectrum sensing algorithms are designed for the single SU case and the case with multiple SUs, respectively (Qin et al. 2014, 2016b). The proposed algorithms contain two phases. In the case with single SU, where signal recovery is to be performed at the SU locally, a new wideband channel division scheme is proposed to reduce the computational complexity of signal recovery in the first phase. In the second phase, a denoising algorithm is performed to improve detection performance by enabling the compressive spectrum sensing algorithm being more robust to channel noise. For the case with multiple cooperative SUs, where spatial diversity among participating SUs is utilized to improve the sensing performance (Ghasemi and Sousa 2005; Akyildiz et al. 2011), the sparse property of spectral signals can be transformed into a low-rank property (Wang et al. 2012). In the first phase, the proposed wideband channel division scheme is invoked to reduce the costs of signal acquisition at SUs. Subsequently, only the compressed measurements are sent to the FC, which reduces the amount of transmission overhead in CRNs. Matrix completion (MC), as a further development of CS, is invoked at the FC to recover the unsensed channels from the sensed channels. In the second phase, detection performance is further improved by the proposed denoising algorithm. To this end, the proposed robust compressive spectrum sensing algorithm is tested on the real-world signals over TVWS after being validated by the simulated TV signals.

1.1.3   Secure Compressive Spectrum Sensing

Along with improving the robustness, adaption, and reducing the complexity of compressive spectrum sensing algorithm, another challenge for CRNs comes from the malicious users, which will send out dishonest data to degrade system performance. In current CSS networks, all cooperative SUs are assumed to be honest and genuine. However, the existence of malicious users would severely degrade the performance of cooperative spectrum sensing (CSS) networks. Moreover, malicious users can degrade the detection performance heavily in sub-Nyquist-based CSS networks. If part of the compressed measurements are corrupted by malicious users, signal recovery would be unstable at the FC.

In order to guarantee the security of CSS networks, a malicious user detection framework is proposed by invoking the low-rank MC technique (Zhang et al. 2014; Qin et al. 2018). More specifically, with the purpose of improving the detection accuracy and reducing the costs of data acquisitions at SUs, the data corrupted by malicious users are removed during the MC process at the FC. Additionally, in order to avoid requiring any prior information of the CSS networks, a rank order estimation algorithm and a malicious user number estimation strategy are proposed. The proposed framework is tested on the real-world signals over TVWS after being validated by the simulated TV signals. Numerical results show that the proposed malicious user detection framework achieves higher detection accuracy with lower costs of data acquisition at SUs or less number of active SUs.

References

Akyildiz, I. F., Lee, W.-Y., Vuran, M. C., & Mohanty, S. (2006). Next generation/dynamic spectrum access/cognitive radio wireless networks: A survey. *Computer Network, 50,* 2127–2159.

Akyildiz, I. F., Lo, B. F., & Balakrishnan, R. (2011). Cooperative spectrum sensing in cognitive radio networks: A survey. *Physical Communication, 4,* 40–62.

Candes, E. (2006). Compressive sampling. In *Proceedings of the International Congress of Mathematicians, Madrid, Spain* (vol. 3, pp. 1433–1452)

Farhang-Boroujeny, B. (2008). Filter bank spectrum sensing for cognitive radios. *IEEE Transactions on Signal Processing, 56,* 1801–1811.

Federal Communications Commission (FCC). (2008). Second report and order and memorandum opinion and order in matter of unlicensed operation in the TV broadcast bands, additional spectrum for unlicensed devices below 900 MHz and in the 3 GHz band, Document 08-260.

Gao, Y., Qin, Z., Feng, Z., Zhang, Q., Holland, O., & Dohler, M. (2016). Scalable and reliable IoT enabled by dynamic spectrum management for M2M in LTE-A. *IEEE Internet of Things Journal, 3,* 1135–1145.

Ghasemi, A., & Sousa, E. (2005). Collaborative spectrum sensing for opportunistic access in fading environments. In *Proceedings of the IEEE International Symposium on Dynamic Spectrum Access Networks (DYSPAN), Baltimore, MD* (pp. 131–136)

Kolodzy, P., & Avoidance, I. (2002). Spectrum policy task force. *Federal Communications Commission, Washington, DC, Rep. ET Docket.*

Landau, H. (1967). Necessary density conditions for sampling and interpolation of certain entire functions. *Acta Mathematica, 117,* 37–52.

Mitola, J., & Maguire, G. Q. (1999). Cognitive radio: Making software radios more personal. *IEEE Personal Communications, 6,* 13–18.

Qin, Z., Gao, Y., & Parini, C. G. (2016a). Data-assisted low complexity compressive spectrum sensing on real-time signals under sub-Nyquist rate. *IEEE Transactions on Wireless Communications, 15,* 1174–1185.

Qin, Z., Gao, Y., Plumbley, M., & Parini, C. (2014). Efficient compressive spectrum sensing algorithm for M2M devices. In *IEEE Global Conference on Signal and Information Processing (GlobalSIP), Atlanta, GA* (pp. 1170–1174).

Qin, Z., Gao, Y., & Plumbley, M. D. (2018). Malicious user detection based on low-rank matrix completion in wideband spectrum sensing. *IEEE Transactions on Signal Processing, 66,* 5–17.

Qin, Z., Gao, Y., Plumbley, M. D., & Parini, C. G. (2016b). Wideband spectrum sensing on real-time signals at sub-Nyquist sampling rates in single and cooperative multiple nodes. *IEEE Transactions on Signal Processing, 64,* 3106–3117.

Qin, Z., Liu, Y., Gao, Y., Elkashlan, M., & Nallanathan, A. (2017). Wireless powered cognitive radio networks with compressive sensing and matrix completion. *IEEE Transactions on Communications, 65,* 1464–1476.

Qin, Z., Wei, L., Gao, Y., & Parini, C. (2015). Compressive spectrum sensing augmented by geo-location database. In *Proceedings of the International Workshop on Smart Spectrum at IEEE Wireless Communications and Networking Conference (WCNC), New Orleans, LA* (pp. 170–175).

Quan, Z., Cui, S., Sayed, A. H., & Poor, H. V. (2009). Optimal multiband joint detection for spectrum sensing in cognitive radio networks. *IEEE Transactions on Signal Processing, 57,* 1128–1140.

Sun, H., Laurenson, D. I., & Wang, C. X. (2010). Computationally tractable model of energy detection performance over slow fading channels. *IEEE Communications Letters, 14,* 924–926.

Sun, H., Nallanathan, A., Wang, C.-X., & Chen, Y. (2013). Wideband spectrum sensing for cognitive radio networks: A survey. *IEEE Wireless Communications, 20,* 74–81.

Tian, Z., & Giannakis, G. (2007). Compressed sensing for wideband cognitive radios. In *IEEE International Conference on Acoustics, Speech, and Signal Processing, Honolulu, HI (ICASSP)* (pp. 1357–1360).

Treichler, J., Davenport, M., & Baraniuk, R. (2009). Application of compressive sensing to the design of wideband signal acquisition receivers. In *US/Australia Joint Work. Defense Apps of Signal Processing (DASP)* (vol. 5).

UK Office of Communications (Ofcom). (2009). Statement on cognitive access to interleaved spectrum.

UK Office of Communications (Ofcom). (2015). Decision to make the wireless telegraphy (White Space Devices).

Wang, Y., Tian, Z., & Feng, C. (2012). Collecting detection diversity and complexity gains in cooperative spectrum sensing. *IEEE Wireless Communications, 11*, 2876–2883.

Zhang, X., Ma, Y., Gao, Y., & Zhang, W. (2018). Autonomous compressive sensing augmented spectrum sensing. *IEEE Transactions on Vehicular Technology, 67*, 6970–6980.

Zhang, X., Qin, Z., & Gao, Y. (2014). Dynamic adjustment of sparsity upper bound in wideband compressive spectrum sensing. In *Proceedings of the IEEE Global Conference on Signal and Information Processing (GlobalSIP), Atlanta, GA* (pp. 1214–1218).

Chapter 2
Sparse Representation in Wireless Networks

Sparse representation of signals has received extensive attention due to its capacity for efficient signal modeling and related applications. The problem solved by the sparse representation is to search for the most compact representation of a signal in terms of a linear combination of the atoms in an overcomplete dictionary. In the literature, three aspects of research on the sparse representation have been focused:

1. Pursuit methods for solving the optimization problem, such as matching pursuit and basis pursuit;
2. Design of the dictionary, such as the K-SVD method;
3. Applications of the sparse representation, such as wideband spectrum sensing, channel estimation of massive MIMO, and data collection in WSNs.

General sparse representation methods, such as principal component analysis (PCA) and independent component analysis (ICA), aim to obtain a representation that enables sufficient reconstruction. It has been demonstrated that PCA and ICA are able to deal with signal corruption, such as noise, missing data, and outliers. For sparse signals without measurement noise, CS can recover the sparse signals exactly with random measurements. Furthermore, the random measurements significantly outperform measurements based on PCA and ICA for the sparse signals without corruption (Chang et al. 2009b; Wright et al. 2010; Chang et al. 2009a).

2.1 Principles of Standard Compressive Sensing

The principles of standard CS, such as to be performed at a single node, can be summarized in the following three parts (Candes 2006).

2.1.1 Sparse Representation

Generally speaking, sparse signals contain much less information than their ambient dimension suggests. Sparsity of a signal is defined as the number of non-zero elements in the signal under a certain domain. Let $\mathbf{f}$ be an N-dimensional signal of interest, which is sparse over the orthonormal transformation basis matrix $\mathbf{\Psi} \in \mathbb{R}^{N \times N}$, and $\mathbf{s}$ be the sparse representation of $\mathbf{f}$ over the basis $\mathbf{\Psi}$. Then $\mathbf{f}$ can be given by

$$\mathbf{f} = \mathbf{\Psi}\mathbf{s}. \tag{2.1}$$

Apparently, $\mathbf{f}$ can be the time or space domain representation of a signal, and $\mathbf{s}$ is the equivalent representation of $\mathbf{f}$ in the $\mathbf{\Psi}$ domain. For example, if $\mathbf{\Psi}$ is the inverse Fourier transform (FT) matrix, then $\mathbf{s}$ can be regarded as the frequency domain representation of the time domain signal, $\mathbf{f}$. Signal $\mathbf{f}$ is said to be K-sparse in the $\mathbf{\Psi}$ domain if there are only K ($K \ll N$) out of the N coefficients in $\mathbf{s}$ that are non-zero. If a signal is able to be sparsely represented in a certain domain, the CS technique can be invoked to take only a few linear and non-adaptive measurements.

2.1.2 Projection

When the original signal f arrives at the receiver, it is processed by the measurement matrix $\mathbf{\Phi} \in \mathbb{R}^{P \times N}$ with $P < N$, to get the compressed version of the signal, that is,

$$\mathbf{x} = \mathbf{\Phi}\mathbf{f} = \mathbf{\Phi}\mathbf{\Psi}\mathbf{s} = \mathbf{\Theta}\mathbf{s}, \tag{2.2}$$

where $\mathbf{\Theta} = \mathbf{\Phi}\mathbf{\Psi}$ is a $P \times N$ matrix, called the sensing matrix. As $\mathbf{\Phi}$ is independent of signal $\mathbf{f}$, the projection process is non-adaptive.

Figure 2.1 illustrates how the different sensing matrices $\mathbf{\Theta}$ influence the projection of a signal from high dimension to its space, i.e., mapping $\mathbf{s} \in \mathbb{R}^3$ to $\mathbf{x} \in \mathbb{R}^2$. As shown in Fig. 2.1, $\mathbf{s} = \begin{pmatrix} s & s & 0 \end{pmatrix}$ is a three-dimensional signal. When $\mathbf{s}$ is mapped into a two-dimensional space by taking $\mathbf{\Theta}_1 = \begin{pmatrix} 1 & -1 & 0 \\ 0 & 0 & 1 \end{pmatrix}$ as the sensing matrix, the original signal $\mathbf{s}$ cannot be recorded based on the projection under $\mathbf{\Theta}_1$. This is because that the plane spanned by the two row vectors of $\mathbf{\Theta}_1$ is orthogonal to signal $\mathbf{s}$ as shown in Fig. 2.1a. Therefore, $\mathbf{\Theta}_1$ corresponds to the worst projection. As shown in Fig. 2.1b, we can also observe that the projection by taking $\mathbf{\Theta}_2 = \begin{pmatrix} 1 & 0 & 0 \\ 0 & 0 & 1 \end{pmatrix}$ is not a good one. It is noted that the plane spanned by the two row vectors of $\mathbf{\Theta}_2$ can only contain part of information of the sparse signal $\mathbf{s}$, and the sparse component in the direction of s_2 is missed when the signal $\mathbf{s}$ is projected into the two-dimensional

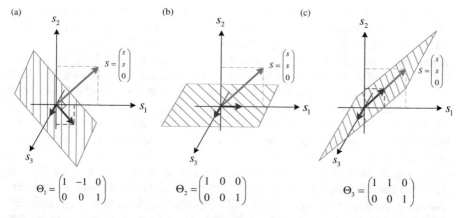

Fig. 2.1 Projection of a sparse signal with one non-zero component with different sensing matrices Qin et al. (2018). (**a**) Worst projection. (**b**) Bad projection. (**c**) Good projection

space. When the sensing matrix is set to $\Theta_3 = \begin{pmatrix} 1 & 1 & 0 \\ 0 & 0 & 1 \end{pmatrix}$, as shown in Fig. 2.1c, the signal **s** can be fully recorded as it falls into the plane spanned by the two row vectors of Θ_3. Therefore, Θ_3 results in a good projection and **s** can be exactly recovered by its projection **x** in the two-dimensional space. Then it is natural to ask what type of projection is good enough to guarantee the exact signal recovery?

The key of CS theory is to find out a stable basis $\boldsymbol{\Psi}$ or measurement matrix $\boldsymbol{\Phi}$ to achieve exact recovery of the signal with length N from P measurements. It seems an undetermined problem as $P < N$. However, it has been proved in Candes et al. (2006) that exact recovery can be guaranteed under the following conditions:

- Restricted isometry property (RIP): Measurement matrix $\boldsymbol{\Phi}$ has the RIP of order K if

$$1 - \delta_K \leq \frac{\|\boldsymbol{\Phi}\mathbf{f}\|_{\ell_2}^2}{\|\mathbf{f}\|_{\ell_2}^2} \leq 1 + \delta_K \tag{2.3}$$

holds for all K-sparse signal $\mathbf{f}$, where δ_K is the restricted isometry constant of a matrix $\boldsymbol{\Phi}$.
- Incoherence property: Incoherence property requires that the rows of measurement matrix $\boldsymbol{\Phi}$ cannot sparsely represent the columns of the sparsifying matrix $\boldsymbol{\Psi}$ and vice versa. More specifically, a good measurement will pick up a little bit information of each component in **s** based on the condition that $\boldsymbol{\Phi}$ is incoherent with $\boldsymbol{\Psi}$. As a result, the extracted information can be maximized by using the minimal number of measurements.

It has been pointed out that verifying both the RIP condition and incoherence property is computationally complicated but they could be achieved with a high probability simply by selecting $\boldsymbol{\Phi}$ as a random matrix. The common random matrices include Gaussian matrix, Bernoulli matrix, or almost all others matrices with independent and identically distributed (i.i.d.) entries. Besides, with the properties of the matrix with i.i.d. entries $\boldsymbol{\Phi}$, the matrix $\boldsymbol{\Theta} = \boldsymbol{\Phi}\boldsymbol{\Psi}$ is also random i.i.d., regardless of the choice of $\boldsymbol{\Psi}$. Therefore, the random matrices are universal as they are random enough to be incoherent with any fixed basis. It has been demonstrated that random measurements can universally capture the information relevant for many compressive signal processing applications without any prior knowledge of either the signal class and its sparse domain or the ultimate signal processing task.

Moreover, for Gaussian matrices the number of measurements required to guarantee the exact signal recovery is almost minimal. However, random matrices inherently have two major drawbacks in practical applications: huge memory buffering for storage of matrix elements, and high computational complexity due to their completely unstructured nature (Candes and Romberg 2007). Compared to the standard CS that limits its scope to standard discrete-to-discrete measurement architectures using random measurement matrices and signal models based on standard sparsity, more structured sensing architectures, named structured CS, have been proposed to implement CS on feasible acquisition hardware. So far, many efforts have been put on the design of structured CS matrices, i.e., random demodulator (Tropp et al. 2010), to make CS implementable with expense of performance degradation. Particularly, the main principle of random demodulator is to multiply the input signal with a high-rate pseudonoise sequence, which spreads the signal across the entire spectrum. Then a low-pass anti-aliasing filter is applied and the signal is captured by sampling it at a relatively low rate. With the additional digital processing to reduce the burden on the analog hardware, random demodulator bypasses the need for a high-rate analog-to-digital converter (ADC) (Tropp et al. 2010). A comparison of Gaussian sampling matrix and random demodulator is provided in Fig. 2.2 in terms of detection probability with different compression ratios P/N. From the figure, the Gaussian sampling matrix performs better than the random demodulator.

2.1.3 Signal Reconstruction

After the compressed measurements are collected, the original signal should be reconstructed. Since most of the basis coefficients in $\mathbf{s}$ are negligible, the original signal can be reconstructed by finding out the minimal set of coefficients that matches the set of compressed measurements $\mathbf{x}$, that is, by solving

$$\hat{\mathbf{s}} = \arg\min_{s} \|s\|_{\ell_p} \text{ subject to } \boldsymbol{\Theta}s = \mathbf{x}, \tag{2.4}$$

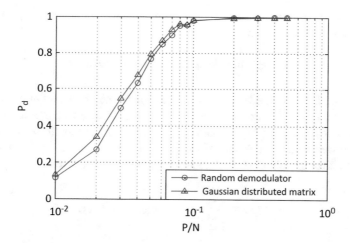

Fig. 2.2 Detection probability versus compression ratio with different measurement matrices. In this case, the signal is one-sparse

where $\|\cdot\|_{\ell_p}$ is the ℓ_p-norm and $p = 0$ corresponds to counting the number of non-zero elements in **s**. However, the reconstruction problem in (2.4) is both numerically unstable and NP-hard (Candes 2006) when ℓ_0-norm is used.

So far, there are mainly two types of relaxations to problem (2.4) to find a sparse solution. The first type is convex relaxation, where ℓ_1-norm is used to substitute ℓ_0-norm in (2.4). Then (2.4) can be solved by standard convex solvers, e.g., cvx. It has been proved that ℓ_1 norm results in the same solution as ℓ_0 norm when RIP is satisfied with the constant $\delta_{2k} < \sqrt{2} - 1$ (Cands 2008). Another type of solution is to use a greedy algorithm, such as OMP (Tropp and Gilbert 2007), to find a local optimum in each iteration. In comparison with the convex relaxation, the greedy algorithm usually requires lower computational complexity and time cost, which makes it more practical for wireless communication systems. Furthermore, the recent result has shown that the recovery accuracy achieved by some greedy algorithms is comparable to the convex relaxation but requiring much lower computational cost (Choi et al. 2017).

2.2 Reweighted Compressive Sensing

As aforementioned, ℓ_1-norm is a good approximation for the NP-hard ℓ_0-norm problem when RIP holds. However, the large coefficients are penalized more heavily than the small ones in ℓ_1-norm minimization, which leads to performance degradation on signal recovery. To balance the penalty on the large and the small coefficients, reweighted CS is introduced by providing different penalties on those large and small coefficients. A reweighted ℓ_1-norm minimization framework

(Candes 2006) has been developed to enhance the signal recovery performance with fewer compressed measurements by solving

$$\hat{\mathbf{s}} = \arg \min_{\mathbf{s}} \|\mathbf{W}\mathbf{s}\|_{\ell_1} \quad \text{subject to} \quad \boldsymbol{\Theta}\mathbf{s} = \mathbf{x}, \tag{2.5}$$

where W is a diagonal matrix with $w_1, \ldots, w_n$ on the diagonal and zeros elsewhere.

Moreover, ℓ_p-norm, e.g., $0 < p < 1$, is utilized to lower the computational complexity of signal recovery process caused by the ℓ_1-norm optimization problem. Iterative reweighted least square (IRLS)-based CS approach has been proposed in Rao and Kreutz-Delgado (1999) to solve (2.4) in a non-convex approach as

$$\hat{\mathbf{s}} = \arg \min_{\mathbf{s}} \sum_{i=1}^{N} w_i s_i \quad \text{subject to} \quad \boldsymbol{\Theta}\mathbf{s} = \mathbf{x}, \tag{2.6}$$

where $w_i = \left| s_i^{(l-1)} \right|^{p-2}$ is computed based on the result of the last iteration, $s_i^{(l-1)}$.

It is worth noting that (2.4) becomes non-convex when $p < 1$. The existing algorithms cannot guarantee to reach a global optimum and may only produce local minima. However, it has been proved (Chartrand 2007; Chartrand and Staneva 2008) that under some circumstances the reconstruction in (2.4) will reach a *unique* and *global* minimizer (Chartrand and Yin 2008), which is exactly $\hat{\mathbf{s}} = \mathbf{s}$. Therefore, we can still exactly recover the signal in practice.

2.3 Distributed Compressive Sensing

The distributed compressive sensing (DCS) (Baron et al. 2009) is an extension of the standard one by considering networks with M nodes. At the m-th node, measurement x_m can be given by

$$x_m = \boldsymbol{\Theta}_m s_m, \quad \forall m \in \mathcal{M}, \tag{2.7}$$

where $\mathcal{M}$ is the set of nodes in the network. As stated in (2.2), $\boldsymbol{\Theta}_m$ is the sensing matrix deployed at the m-th node, and s_m is a sparse signal of interest. DCS becomes a standard CS when $M = 1$.

In the applications of standard CS, the signal received at the same node has its sparsity property due to its intra-correlation. While for the networks with multiple nodes, signals received at different nodes exhibit strong inter-correlation. The intra-correlation and inter-correlation of signals from the multiple nodes lead to a joint sparsity property. The joint sparsity level is usually smaller than the aggregate over the individual signal's sparsity level. As a result, the number of compressed measurements required for exact recovery in DCS can be reduced significantly compared to the case performing standard CS at each single node independently.

In DCS, there are two closely related concepts: distributed networks and distributed CS solvers. The distributed networks refer to networks that different nodes perform data acquisition in a distributed way and the standard CS can be applied at each node individually to perform signal recovery. While for DCS solver as proposed in Baron et al. (2009), the data acquisition process requires no collaboration among sensors and the signal recovery process is performed at several computational nodes, which can be distributed in a network or locally placed within a multiple core processor. Generally, it is of interest to minimize both computation cost and communication overhead in DCS. The most popular application scenario of DCS is that all signals share the common sparse support but with different non-zero coefficients.

2.4 Compressive Spectrum Sensing

The last decade has witnessed the rapid explosion of wireless devices all over the world, which gives rise to the increasing demand for wireless spectral resource. As reported by FCC and Ofcom (Kolodzy and Avoidance 2002; Ofc a), there are significant temporal and spatial variations in the allocated spectrum. Given this fact, CR has been proposed as an intelligent system to detect spectrum holes for unlicensed usage (Mitola and Maguire 1999). More specifically, the basic idea of CR is to match the requirements of higher layer applications or users with the available resources. The available resources include available power, spectrum, and other resources that can be utilized by unlicensed SUs. CR is a radio that is capable of sensing the available resources and learning from the user behaviors and its previous decisions and mistakes, in order to provide a better response to the new resource request from SUs. So far, CR has been widely investigated.

2.4.1 Spectrum Sensing Methods

In CR, spectrum sensing is one of the most challenging tasks, which allows SUs to have the knowledge of spectrum occupancy. Once a spectrum hole is detected, SUs can make use of it for data transmission. Spectrum sensing requires high accuracy and low complexity for DSA (Nekovee 2008). There is an extensive research work on spectrum sensing techniques being carried out. Many theoretical models for spectrum sensing techniques have been proposed, such as matched filter detection, cyclostationary feature detection, and energy detection. The matched filter detection is an optimal detection method that requires the prior information of PUs (Bhargavi and Murthy 2010). However, it requires SUs to have a dedicated sensing receiver for each type of PU signals. Cyclostationary feature detection can distinguish the PUs and noise by utilizing the periodicity in the received primary signal. However, it requires high computational complexity and prior information of the primary signals. Among these three approaches for spectrum sensing, energy detection is a non-coherent detection method, which avoids the requirement for prior knowledge of PUs. Additionally, energy detection approach does not require complicated receivers as the other two approaches do. Therefore, it is easy to be implemented, and the computational complexity is relatively low, but with a drawback of poor detection performance under low SNR scenarios. In this book, energy detection is adopted due to its simplicity.

2.4.2   Spectrum Sensing Model

In spectrum sensing, the received signal can be expressed as

$$r(t) = s(t) + w(t), \tag{2.8}$$

where $s(t)$ is the signal to be detected, $w(t)$ is the additive white Gaussian noise (AWGN) samples with noise variance σ_n^2. It is noted that $s(t) = 0$ when there is no transmission by PU. The energy of sensed signals can be written as

$$E = \int_0^T |r(t)|^2 dt, \tag{2.9}$$

where T is the sensing period. When Nyquist sampling is performed, N samples are collected during one sensing period T. The decision on spectrum occupancy can be obtained by comparing the energy E of the received signal with a threshold λ. Particularly, the sensing decision can be formulated into a binary hypothesis problem

$$\begin{aligned} \mathcal{H}_0 &: r(t) = w(t), \\ \mathcal{H}_1 &: r(t) = s(t) + w(t), \end{aligned} \tag{2.10}$$

where $\mathcal{H}_0$ and $\mathcal{H}_1$ denote the hypothesis that PU is absent and present, respectively.

Additionally, the performance of energy detection algorithm can be measured by two probabilities: probability of detection P_d and probability of false alarm P_f. P_d is the probability of detecting a signal on the considered frequency when it actually is present. P_f is the probability that the test incorrectly decides that the considered frequency is occupied when it actually is not. With a target probability of false alarm $\bar{P}_f$, the threshold λ is given by

$$\lambda = \sigma_n^2 \left(1 + \frac{Q^{-1}\left(\bar{P}_f\right)}{\sqrt{N/2}} \right). \tag{2.11}$$

If the target probability of detection $\bar{P}_d$ is given, the threshold can be calculated as

$$\lambda = \left(\sigma_s^2 + \sigma_n^2\right)\left(1 + \frac{Q^{-1}\left(\bar{P}_d\right)}{\sqrt{N/2}}\right), \tag{2.12}$$

where σ_s^2 refers to the power of the transmitted primary signal.

2.4.3   Compressive Wideband Spectrum Sensing

Inspiring by the most recent developments on CS and MC techniques, the bottleneck of Nyquist wideband sensing in CRNs can be broken through compressive spectrum sensing. The compressive spectrum sensing at a single SU is taken as an example. In the considered model, it is assumed that bandwidth of the whole spectrum is divided into I channels. A channel is either occupied by a PU or unoccupied. Meanwhile, there is no overlap between different channels. The number of occupied channels K is assumed to be much less than the total number of channels I. As shown in Fig. 2.3, the compressive spectrum sensing model includes the following four steps:

1. Signal arrives at SUs;
2. Compressed measurements collection at SUs;
3. Signal recovery;
4. Decision making for spectrum occupancy.

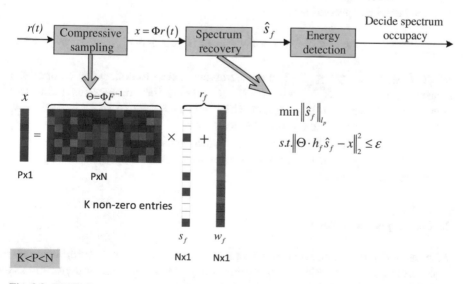

Fig. 2.3 Block diagram for compressive spectrum sensing model

2.4.3.1 Signals Arrives at Secondary Users

The signals transmitting over the spectrum of interest are defined as $s(t) \in \mathscr{C}^{N \times 1}$, where N is the number of samples when $s(t)$ is sampled at or above Nyquist rate. Consequently, the signals received at an SU are given by

$$r(t) = h(t) * s(t) + w(t), \qquad (2.13)$$

where $h(t)$ is the channel gain between the transmitter and receiver, and $w(t) \sim \mathscr{CN}(0, \sigma_n^2 \mathbf{I}_N)$ refers to AWGN. Here, σ_n^2 refers to noise variance, and $\mathbf{I}_N$ is the identity matrix.

In order to make sure CS technique working well at SUs, the received signal $r(t)$ should be able to be expressed in a sparse domain. In spectrum sensing, as shown in Fig. 2.3, the signals $r(t)$ received at an SU are assumed to be sparse in the frequency domain, as the spectrum utilization is low in reality. Here, the sparse representation of the received signal can be expressed as

$$\mathbf{r_f} = \mathbf{h_f s_f} + \mathbf{w_f}, \qquad (2.14)$$

where $\mathbf{r_f}$, $\mathbf{h_f}$, $\mathbf{s_f}$, and $\mathbf{w_f}$ refer to the discrete Fourier transform (DFT) of $r(t)$, $h(t)$, $s(t)$, and $w(t)$, respectively.

2.4.3.2 Compressed Measurements Collection

After the CS technique is invoked at an SU, the compressed measurements collected at the SU can be expressed as:

$$\mathbf{x} = \boldsymbol{\Phi} \mathscr{F}^{-1} \mathbf{r}_f = \boldsymbol{\Theta} \mathbf{r}_f = \boldsymbol{\Theta} \left(h_f s_f + w_f \right), \qquad (2.15)$$

where $\boldsymbol{\Phi} \in \mathscr{C}^{P \times N}$ $(P \leq N)$ is a measurement matrix to collect the compressed measurements $\mathbf{x} \in \mathscr{C}^{P \times 1}$, with $P/N \leq 1$ being the compression ratio. The measurement matrix can be a matrix which contains a single spike in each row. Then the case $P/N = 1$ corresponds to $\boldsymbol{\Phi} = \mathbf{I}_N$. Additionally, $\boldsymbol{\Theta} = \boldsymbol{\Phi} \mathscr{F}^{-1}$, where $\mathscr{F}^{-1}$ is inverse DFT (IDFT) matrix which is used as the sparsifying matrix. In practical settings, structured random matrices are often employed for improved implementation affordability.

2.4.3.3 Signal Recovery

With the available compressed measurements, the original signal should be recovered before making decision on spectrum occupancy. As l_0-norm minimization is an NP-hard problem, it has been proved in Candes (2006) that l_1-norm is a good approximation for l_0-norm. Taking the l_1-norm minimization as the CS solver, signal recovery can be performed at an SU by solving the following convex optimization problem:

$$\min \|\hat{s}_f\|_1,$$
$$\text{subject to} \quad \|\boldsymbol{\Theta} \cdot h_f \hat{s}_f - \boldsymbol{x}\|_2^2 \leq \varepsilon, \tag{2.16}$$

where $\hat{s}_f$ refers to the recovered signal, and ε refers to the noise tolerance. In the case of CS-based CSS, each participating SU sends the compressed measurements to the FC, which is the place to perform signal recovery.

2.4.3.4 Decision Making

When the reconstructed signal $\hat{s}_f$ is obtained, energy detection can be performed to determine the spectrum occupancy. More specifically, the energy density of each recovered channel is compared with a predefined threshold to determine whether the corresponding channel is occupied or not. The predefined threshold λ is as defined in (2.11) or (2.12), which is dependent on whether a target P_d or P_f is given.

In practice, the noise variance σ_n^2 can be calibrated in a given channel, which is known for sure to be idle. For example, some channels, such as channel 21 in TVWS, are supposed to be vacant currently in the UK Ofc (a). If the energy density of the considered channel is higher than the threshold, the corresponding channel is determined as occupied by PUs. Consequently, SUs are forbidden to access it. Otherwise, the corresponding channel is determined as vacant. Therefore, SUs can access it to transmit the unlicensed signals.

2.5 Summary

This chapter presents the fundamental concept of CS as well as frameworks of CS-based wideband spectrum sensing.

References

Baron, D., Duarte, M. F., Sarvotham, S., Wakin, M. B., & Baraniuk, R. G. (2009). Distributed compressive sensing. arXiv:0901.3403, https://arxiv.org/abs/0901.3403.

Bhargavi, D., & Murthy, C. R. (2010). Performance comparison of energy, matched-filter and cyclostationarity-based spectrum sensing. In *International Workshop on Signal Processing Advances in Wireless Communications (SPAWC)* (pp. 1–5).

Candes, E. (2006). Compressive sampling. In *International Congress of Mathematicians, Madrid, Spain* (vol. 3, pp. 1433–1452).

Candes, E., & Romberg, J. (2007). Sparsity and incoherence in compressive sampling. *Inverse Problems, 23*, 969.

Cands, E. J. (2008). The restricted isometry property and its implications for compressed sensing. *Comptes Rendus Mathematique, 346*, 589–592.

Candes, E. J., Romberg, J., & Tao, T. (2006). Robust uncertainty principles: Exact signal reconstruction from highly incomplete frequency information. *IEEE Transactions on Information Theory, 52*, 489–509.

Chang, H. S., Weiss, Y., & Freeman, W. T. (2009a). Informative sensing. CoRR abs/0901.4275.

Chang, H. S., Weiss, Y., & Freeman, W. T. (2009b). Informative sensing of natural images. In *IEEE International Conference on Image Processing (ICIP), Cairo, Egypt* (pp. 3025–3028).

Chartrand, R. (2007). Exact reconstruction of sparse signals via nonconvex minimization. *IEEE Signal Processing Letters, 14*, 707–710.

Chartrand, R., & Staneva, V. (2008). Restricted isometry properties and nonconvex compressive sensing. *Inverse Problems, 24*, 035020.

Chartrand, R., & Yin, W. (2008). Iteratively reweighted algorithms for compressive sensing. In *Proceedings of the IEEE International Conference on Acoustics, Speech, and Signal Processing (ICASSP), Las Vegas, NV* (pp. 3869–3872).

Choi, J. W., Shim, B., Ding, Y., Rao, B., & Kim, D. I. (2017). Compressed sensing for wireless communications: Useful tips and tricks. *IEEE Communications Surveys & Tutorials, 19*, 1527–1550.

Kolodzy, P., & Avoidance, I. (2002). Spectrum policy task force. Federal Communications Commission, Washington, DC, Rep. ET Docket.

Mitola, J., & Maguire, G. Q. (1999). Cognitive radio: Making software radios more personal. *IEEE Personal Communications, 6*, 13–18.

Nekovee, M. (2008). Impact of cognitive radio on future management of spectrum. In *International Conference on Cognitive Radio Oriented Wireless Networks and Communications (Crown-Com), Singapore* (pp. 1–6)

Qin, Z., Fan, J., Liu, Y., Gao, Y., and Li, G. Y. (2018). Sparse representation for wireless communications: A compressive sensing approach. *IEEE Signal Processing Magazine, 35*, 40–58.

Rao, B. D., & Kreutz-Delgado, K. (1999). An affine scaling methodology for best basis selection. *IEEE Transactions on Signal Processing, 47*, 187–200.

Tropp, J., Laska, J., Duarte, M., Romberg, J., & Baraniuk, R. (2010). Beyond Nyquist: Efficient sampling of sparse bandlimited signals. *IEEE Transactions on Information Theory, 56*, 520–544.

Tropp, J. A., & Gilbert, A. C. (2007). Signal recovery from random measurements via orthogonal matching pursuit. *IEEE Transactions on Information Theory, 53*, 4655–4666.

UK Office of Communications (Ofcom). (2009). Statement on cognitive access to interleaved spectrum.

Wright, J., Ma, Y., Mairal, J., Sapiro, G., Huang, T. S., & Yan, S. (2010). Sparse representation for computer vision and pattern recognition. *Proceedings of IEEE, 98*, 1031–1044.

Part II
Compressive Spectrum Sensing Algorithms

Chapter 3
Data-Driven Compressive Spectrum Sensing

In this chapter, the related work and the main contributions are firstly introduced in Sect. 3.1. In Sect. 3.2, the proposed data-driven compressive spectrum sensing framework is presented, in which geolocation database is used to provide prior information for signal recovery. Additionally, Sect. 3.3 gives the numerical results of the proposed framework. Finally, Sect. 3.4 concludes this chapter.

3.1 Introduction

In order to avoid any harmful interference to primary services in TVWS, SUs, also named as WSDs, should have the knowledge of spectrum occupancy. Two approaches have been proposed to make SUs aware of the spectrum occupancy. One approach is geolocation database which is a centralized database to output the maximum allowable EIRP for each vacant TVWS channel for a specific location and time (Kolodzy and Avoidance 2002). Geolocation database typically calculates the interference generated in wireless communication systems through theoretical propagation models rather than actual measurements, which may result in inaccurate results for spectrum occupancy (Paisana et al. 2014). Furthermore, geolocation database approach can only protect the registered users. However, some SUs may not be registered, which may pose significant challenges to a geolocation database. For example, PMSE devices such as wireless microphone operate mostly on an unlicensed basis, without any record in TVWS (Ribeiro et al. 2012). The approach to protect unregistered applications is spectrum sensing. Spectrum sensing requires SUs to have the capability to detect spectrum holes that are not occupied by PUs. This approach provides instant channel occupancy information, but it may cause interference to some reserved channels which would be determined as vacant by sensing only. Therefore, a geolocation database can be utilized to improve the accuracy of spectrum sensing.

© The Author(s), under exclusive license to Springer Nature Switzerland AG 2019
Y. Gao, Z. Qin, *Data-Driven Wireless Networks*, SpringerBriefs in Electrical
and Computer Engineering, https://doi.org/10.1007/978-3-030-00290-9_3

3.1.1   Related Work

So far, some work has been researched on the combination of spectrum sensing and geolocation database. Wang et al. (2014) proposed a framework combining spectrum sensing with geolocation database was proposed, in which the utilization of spatial-temporal spectrum hole is maximized. Wang et al. (2015) proposed to combine the advantages of spectrum sensing and geolocation database, in which different spectrum sensing modules are performed based on the output of geolocation database. Furthermore, Ribeiro et al. (2012) implemented a framework into an experimental platform by combining wireless microphone sensors with a web-based geolocation database access for PMSE. However, all the existing frameworks required that SUs should build a direct link to the remote geolocation database. This direct link causes increasing loads in CR networks.

Besides the work on framework combining spectrum sensing and geolocation database, wideband spectrum sensing has attracted much attention. As limited by the Nyquist sampling theory, CS has been proposed to achieve sub-Nyquist rate by utilizing the natural sparse property of signals (Candes 2006). So far, amount of work has been done on compressive spectrum sensing (Tian and Giannakis 2007; Wang et al. 2012; Sun et al. 2012). Many of the existing algorithms utilize l_1-norm minimization. However, as pointed out in Candes et al. (2008), large coefficients are penalized more heavily than smaller coefficients in l_1 minimization, which may lead to performance degeneration. In order to rectify a key difference between l_0 and l_1 minimization and balance the penalty on large coefficients and smaller coefficients of the sparse signal, Candes et al. (2008) proposed an iteratively reweighted l_1 minimization algorithm by introducing weight for each bin of the signal to be recovered. Another approach to recover a sparse signal with fewer measurements is to replace the l_1 norm with l_p norm. In order to solve the l_p norm problem, an IRLS algorithm was proposed to perform sparse signal reconstruction (Chartrand 2007; Chartrand and Yin 2008; Chartrand and Staneva 2008; Carrillo and Barner 2009; Saab and Yılmaz 2010; Ba et al. 2014).

Moreover, recovering signals from compressed measurements by utilizing prior information has been studied in Escoda et al. (2006), Friedlander et al. (2012), Lu and Vaswani (2012) and Miosso et al. (2009). Specifically, Escoda et al. (2006) proposed the prior information assisted sparse signal approximation algorithms: weighted basis pursuit denoising and weighted match pursuit. Additionally, two partial support information assisted CS algorithms were proposed respectively in Friedlander et al. (2012) and Lu and Vaswani (2012), in which the weighted l_1 minimization approach with fixed weights on the known support is utilized to find the sparse solution for CS problems. Furthermore, Miosso et al. (2009) proposed an IRLS-based CS recovery algorithm utilizing the prior information, in which the weights are updated in each iteration of the IRLS algorithm. The different iterative approaches for weight setting in IRLS were compared in Wipf and Nagarajan (2010). However, the iterative weight updating approach in IRLS introduces extra computational complexities for signal recovery.

3.1.2 Contributions

Motivated by the challenges identified above, the main contributions of this chapter are listed as follows:

1. A data-driven compressive spectrum sensing framework is proposed, in which a geolocation database algorithm is implemented at SUs locally to provide prior information for the compressive spectrum sensing.
2. In the proposed framework, a DNRLS-based compressive spectrum sensing algorithm with lower computational complexity and fewer compressed measurements is proposed. In the proposed DNRLS, data generated by the locally stored geolocation database algorithm is utilized to replace the iterative process of weights updating in IRLS algorithm. Convergence and computational complexity of the proposed DNRLS are analyzed.
3. Additionally, an efficient approach for calculating the maximum allowable EIRP is proposed to further improve the accuracy and efficiency of the geolocation database algorithm stored at SUs.
4. Furthermore, based on recent work on the trial within the Ofcom TVWS pilot (Holland et al. 2015), the proposed framework and algorithms are tested on real-world signals and data after being validated by the simulated signals and data.

3.2 Data-Driven Compressive Spectrum Sensing Framework

In the wideband spectrum sensing scenario, as shown in Fig. 3.1a, multiple PUs exist in the multiband spectrum of interest and each SU is capable to detect the active PUs accurately and efficiently. The traditional hybrid frameworks with geolocation database and spectrum sensing proposed in Ribeiro et al. (2012), Wang et al. (2014) and Wang et al. (2015) require a direct link to the remote geolocation database as shown in Fig. 3.1b. Dynamic changes of the spectrum would not be reflected unless the users are registered and updated in the centralized geolocation database. This process introduces several information exchanges such as the two-way transmissions between the SU and the geolocation database. Additionally, each transmission link introduces extra energy consumption at SUs and requires bandwidth for information exchange.

In order to reduce the necessary sampling rates at SUs and alleviate both the network load and the transmission errors between geolocation database and SUs, a framework is proposed by combining compressive spectrum sensing with geolocation database algorithm, which is named as DNRLS framework as shown in Fig. 3.1c. In the proposed framework, the DTT database is maintained at the SU locally, which includes the DTT transmitter (TV base station) information. The geolocation database calculation algorithm is employed locally at the SU to calculate the maximum allowable EIRP P_{IB} of each TV channel based on the

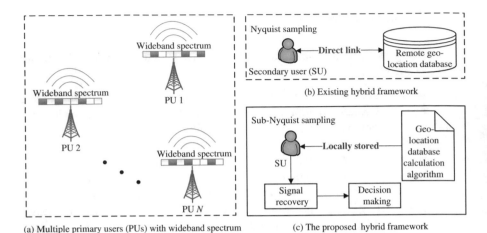

(a) Multiple primary users (PUs) with wideband spectrum   (c) The proposed hybrid framework

Fig. 3.1 (**a**) Scenario of wideband spectrum sensing with multiple primary users (PUs); (**b**) the existing framework with a conventional spectrum sensing at Nyquist rate and a direct link to remote database; and (**c**) the proposed DNRLS framework

DTT database and geography location information of the SU. Before starting a new sensing period, the SU firstly collects its own geography location information by GPS. Then the location information is utilized as the input of geolocation database calculation algorithm to calculate P_{IB} of each TV channel at the SU locally. Subsequently, the obtained P_{IB} of each channel is mapped to the instant spectrum occupancy information and then fused with the historical spectrum occupancy information. The fused results can provide an estimation on the sparsity level of the spectrum of interest. According to the estimated sparsity level, SU can determine the minimal sampling rate to collect compressed measurements that can guarantee exact signal recovery. After the compressed measurements are obtained, fused channel occupancy information is utilized as the prior information for solving signal recovery problem for compressive spectrum sensing. As a result, necessary sampling rates for exact signal recovery and computational complexities are reduced at SUs. After the original signal is recovered, the decision on spectrum occupancy can be made by employing energy detection method. Furthermore, in order to further relax the SU, a Wilkinson's method (Fenton 1960) is adopted to calculate the maximum allowable EIRP P_{IB} of each TV channel efficiently.

3.2.1 Iteratively Reweighted Least Square-Based Compressive Sensing

Before introducing the proposed DNRLS-based compressive spectrum sensing algorithm, the IRLS algorithm is introduced. As aforementioned, l_1-norm has been proved as a good approximation for the NP-hard l_0-norm problem. However, as pointed out in Candes et al. (2008), large coefficients are penalized more heavily than smaller coefficients in l_1 minimization, which may lead to performance degeneration. In order to balance the penalty on large coefficients and small coefficients in the signal to be recovered, an iterative process to construct the weights is introduced. Additionally, l_p-norm ($0 < p < 1$) is utilized to lower the computational complexity of signal recovery process caused by solving the l_1-norm optimization problem. Furthermore, IRLS-based compressive sensing has been proposed to utilize the l_p-norm to reduce computational complexity of signal recovery (Chartrand 2007; Chartrand and Yin 2008; Chartrand and Staneva 2008; Carrillo and Barner 2009; Saab and Yılmaz 2010; Ba et al. 2014). Meanwhile, the signal recovery performance is improved by introducing the iteratively updated weights.

Based on the compressive spectrum sensing model introduced in Sect. 2.4.3 of Chap. 3, with the IRLS algorithm, the original signal $\hat{\mathbf{s}}_\mathbf{f}$ can be obtained by solving the following problem in Lagrangian form:

$$\min \ \left\| \boldsymbol{\Theta} \cdot \mathbf{h_f}\hat{\mathbf{s}}_\mathbf{f} - \mathbf{x} \right\|_2^2 + \lambda_L \mathbf{W}\hat{\mathbf{s}}_\mathbf{f}^2, \tag{3.1}$$

where $\hat{\mathbf{s}}_\mathbf{f}$ refers to the signal to be reconstructed, and $\mathbf{h_f}$ refers to the related channel coefficients. $\mathbf{W} = \mathrm{diag}\left\{ \frac{1}{w_1}, \ldots, \frac{1}{w_n}, \ldots, \frac{1}{w_N} \right\}$ is a diagonal matrix which is computed from the previous iteration and updated in each iteration. Here, w_n refers to the weight for the sample indexed by n, and $\mathbf{x} \in \mathscr{C}^{P \times 1}$. The compression ratio is defined as $\gamma = \frac{P}{N}$. Additionally, λ_L is the Lagrangian factor. In the lth ($l = 0, 1, \ldots, I_{\max}$) iteration of the IRLS algorithm, the weights are calculated with the recovered signal $\hat{\mathbf{s}}_{\mathbf{f\_n}}^{(l-1)}$ in the $(l-1)$th iteration as

$$w_n^{(l)} = \left(\left(\hat{\mathbf{s}}_{\mathbf{f\_n}}^{(l-1)} \right)^2 + \zeta^{(l)} \right)^{\frac{p}{2}-1}. \tag{3.2}$$

In (3.2), $\zeta^{(l)}$ is updated in each iteration, and it is a positive value to make sure that a zero-valued component in $\hat{\mathbf{s}}_\mathbf{f}$ does not strictly prohibit a non-zero estimate in the next iteration of weights update. Additionally, the solution of (3.1) at the lth iteration can be expressed as

$$\hat{\mathbf{s}}_{\mathbf{f\_n}}^{(l)} = \mathbf{W}^{(l)} \boldsymbol{\Theta}^{\mathrm{T}} \left(\mathbf{h_f} \boldsymbol{\Theta} \mathbf{W}^{(l)} \boldsymbol{\Theta}^{\mathrm{T}} + \lambda_L \mathbf{I}_P \right)^{-1} \mathbf{x}, \tag{3.3}$$

where the initial value for the weights w_n in $\mathbf{W}$ is 1, and then $\mathbf{W}^{(0)} = \mathbf{I}_N$. As a result, $\hat{\mathbf{s}}_{\mathbf{f\_n}}^{(0)} = \boldsymbol{\Theta}^{\mathrm{T}}(\mathbf{h_f} + \lambda_L \mathbf{I}_P)^{-1} x$. It is noted that (3.1) is a convex optimization problem when $p = 1$, and a non-convex optimization problem when $0 < p < 1$. As such, the solution to (3.1) can be local minima when $0 < p < 1$. Even though no theoretic guarantee, the numerical results in Chartrand (2007), Chartrand and

Yin (2008) and Chartrand and Staneva (2008) has shown that the computed local minimizer of (3.1) is the global one when it is solved by IRLS.

Definition 1 The RIP guarantees the stable and robust recovery by solving the optimization problem (2.16). We say that a matrix $\boldsymbol{\Theta}$ satisfies the property (a, K, p) if it satisfies

$$\delta_{aK} + a^{\frac{2}{p}-1}\delta_{(a+1)K} < a^{\frac{2}{p}-1} - 1, \tag{3.4}$$

where $a > 1$, and K is sparsity level of the spectrum of interest.

Theorem 1 Let $0 < p \le 1$. If a $P \times N$ matrix satisfies $P(a, K, p)$, then (Saab and Yılmaz 2010)

$$\left\|\hat{\mathbf{s}}_{\mathbf{f}} - \mathbf{s}_{\mathbf{f}}\right\|_2^p \le C^1 \eta + C^2 \frac{\left\|\mathbf{s}_{\mathbf{f}} - \mathbf{s}_{\mathbf{f},\mathbf{K}}\right\|_p^p}{K^{\frac{2}{p}-1}}, \tag{3.5}$$

where $C^1 = 2^p \dfrac{1 + a^{\left(\frac{p}{2}-1\right)}\left(\frac{p}{2}-1\right)^{-\frac{p}{2}}}{(1-\delta_{(a+1)K})^{\frac{p}{2}} - (1+\delta_{aK})^{\frac{p}{2}} a^{\left(\frac{p}{2}-1\right)}}$, and

$$C^2 = \frac{2\left(\frac{p}{2-p}\right)^{\frac{p}{2}}}{a^{\left(1-\frac{p}{2}\right)}} \left(1 + \frac{\left(\left(\frac{p}{2}-1\right)^{-\frac{p}{2}} + a^{\left(\frac{p}{2}-1\right)}\right)(1+\delta_{aK})^{\frac{p}{2}}}{(1-\delta_{(a+1)K})^{\frac{p}{2}} - (1+\delta_{aK})^{\frac{p}{2}} a^{\left(\frac{p}{2}-1\right)}}\right).$$

3.2.2 Non-iteratively Reweighted Least Square-Based Compressive Sensing

In the traditional IRLS-based CS given by (3.1), the key challenge is to find the optimal set of weights $\mathbf{W}$ in an iterative process for a better estimate of the original signals. It should be noted that the iterations generate more computational complexities during the signal recovery process. When part of the maximum allowable EIRP is available in advance, the iterative process can be removed without degrading the recovery performance heavily. In this chapter, a DNRLS-based compressive spectrum sensing algorithm is proposed. In the proposed algorithm, a geolocation database algorithm is implemented at SUs locally to provide data for weights calculation. It is achieved by a non-iterative method, so that SUs do not need any additional link to a centralized geolocation database. Based on (3.2), the proposed calculation yields the weights as

$$w_n = \left(|\bar{\gamma}_n|^2 + \zeta\right)^{\frac{p}{2}-1}, \tag{3.6}$$

where ζ is a positive value same as $\zeta^{(l)}$ in (3.2), and $\bar{\boldsymbol{\gamma}} = \{\bar{\gamma}_1, \ldots \bar{\gamma}_n, \ldots \bar{\gamma}_N\}$ is constructed by the channel historical data and the output of geolocation database algorithm. By introducing weights to solving the optimization problem (3.1), the samples with high power density will be penalty by relative light weights. While for the samples with low power density, the weighted penalty will be relatively large. By doing so, the optimization result of (3.1) will be more close to the solution of the original l_0-norm problem. The construction of $\bar{\gamma}$ in detail is introduced in the following.

In the $(t + 1)$th sensing period, the maximum allowable EIRP $P_{IB}(t + 1)$ is calculated for the current period by the proposed Wilkinson's method-based DTT location probability calculation algorithm introduced in Sect. 3.2.3. Subsequently, the $P_{IB}(t + 1)$ is mapped to $\gamma(t + 1)$. Furthermore, the averaged $\bar{\gamma}(t + 1)$ is updated by fusing $\bar{\gamma}(t + 1)$ with $\bar{\gamma}(t)$ as

$$\bar{\gamma}(t + 1) = \xi\bar{\gamma}(t) + (1 - \xi)\gamma(t + 1), \tag{3.7}$$

where $\bar{\gamma}(t)$ is the historical data for the weights construction at the tth sensing period with $t = \{0, 1, \ldots, T\}$, and ξ $(0 < \xi < 1)$ is the weight for $\bar{\gamma}(t)$. Herein T is the window size for SUs to fuse the current allowable maximum P_{IB} with the historical data. At an SU, only the $\bar{\gamma}(t)$ is stored locally after the tth sensing period. If there is any new unregistered user showing up in the spectrum of interest in tth period, the related DTT transmitter information used for geolocation database calculation algorithm is updated locally. This makes the proposed weights calculation robust to the new unregistered users. Meanwhile, the geolocation database at other SUs would not be influenced. In the $(t + 1)$th period, the $\gamma(t + 1)$ provided by the local geolocation database calculation algorithm would be updated accordingly by considering the unregistered users. After $\bar{\gamma}(t + 1)$ for the current sensing period is obtained to calculate the weights, a more accurate spectrum estimation can be obtained by solving the following non-iterative problem:

$$\hat{s}_f = \tilde{W}\Theta^T \left(h_f\hat{s}_f\tilde{W}\Theta^T + \lambda_L I_P \right)^{-1} x. \tag{3.8}$$

In (3.8), $\tilde{W} = \text{diag}\left(\frac{1}{w_1}, \ldots, \frac{1}{w_n}, \ldots, \frac{1}{w_1}\right)$ is a diagonal matrix in which w_n is calculated by (3.6) to replace the iterative update process in (3.2). In the proposed DNRLS-based compressive spectrum sensing algorithm, the accuracy of $\bar{\gamma}$ would affect the recovery performance.

3.2.2.1 Convergence Analyses

If there is no unregistered user in the spectrum of interest, which means the values of $\bar{\gamma}$ used to construct the weights are accurate, the recovery performance of DNRLS is very good. When the unregistered users show up in the spectrum of interest at the first sensing period, the $\bar{\gamma}(1)$ becomes inaccurate on the corresponding bins as the output of the local geolocation database algorithm $\gamma(1)$ for the first period is inaccurate. As a result, the signal recovery and detection performance would be degraded accordingly. In the tth period after the unregistered user shows up in the spectrum of interest, $\boldsymbol{\gamma}(t)$ is fused with the historical data $\bar{\boldsymbol{\gamma}}(t-1)$ of the $(t-1)$th period. The accuracy of weights $\bar{\boldsymbol{\gamma}}(T)$ is dependent on the window size T for the weights fusion at SUs. The weights fusion process is shown as follows:

$$\begin{aligned}
\bar{\boldsymbol{\gamma}}(1) &= \xi\bar{\boldsymbol{\gamma}}(0) + (1-\xi)\,\boldsymbol{\gamma}(1)\,, & \text{(1st period)} \\
\bar{\boldsymbol{\gamma}}(1) &= \xi\bar{\boldsymbol{\gamma}}(1) + (1-\xi)\,\boldsymbol{\gamma}(2)\,, & \text{(1st period)} \\
\bar{\boldsymbol{\gamma}}(T) &= \xi\bar{\boldsymbol{\gamma}}(T-1) + (1-\xi)\,\boldsymbol{\gamma}(T)\,, & \text{(Tth period)}
\end{aligned} \tag{3.9}$$

where $\bar{\boldsymbol{\gamma}}(0)$ is the historical data for weights construction before unregistered user showing up, and $\boldsymbol{\gamma}(1)$ is the output of the locally implemented geolocation database algorithm for the period when unregistered users show up in the spectrum of interest. As $\boldsymbol{\gamma}(2) = \cdots = \boldsymbol{\gamma}(T) = \boldsymbol{\gamma}$, which represents the real spectrum status with consideration of the unregistered users in the spectrum of interest, $\bar{\boldsymbol{\gamma}}(T)$ can be expressed as

$$\bar{\boldsymbol{\gamma}}(T) = \xi^T \times \bar{\boldsymbol{\gamma}}(0) + (1-\xi)\,\xi^{T-1} \times \boldsymbol{\gamma}(1) + \frac{(1-\xi) \times \boldsymbol{\gamma} \times \left(1-\xi^{T-1}\right)}{1-\xi}$$

$$\tag{3.10}$$

$$= \xi^T \times \bar{\boldsymbol{\gamma}}(0) + (1-\xi)\,\xi^{T-1} \times \boldsymbol{\gamma}(1) + \left(1 - (\xi)^{T-1}\right) \times \boldsymbol{\gamma}.$$

It is noted that $\bar{\boldsymbol{\gamma}}(T)$ will converge fast to $\boldsymbol{\gamma}$ after unregistered users show up in the spectrum of interest. The smaller ξ, the convergence speed goes faster. Additionally, part of channels in TVWS are fixed and utilized by DTV signals, and some of the channels are reserved for other purposes. As a result, at least the weights for those fixed channels in $\bar{\boldsymbol{\gamma}}(0)$ and $\boldsymbol{\gamma}(1)$ are accurate. This characteristic provides a guarantee that the recovery performance would not be degraded heavily when unregistered users show up in the spectrum of interest. With increasing window size T, the influence of inaccurate weights in $\bar{\boldsymbol{\gamma}}(0)$ and $\boldsymbol{\gamma}(1)$ degrades. The influence of the window size T is shown in the numerical analyses part in Sect. 3.3.

3.2.2.2 Complexity Analyses

The computational complexity reduction of the proposed DNRLS-based compressive spectrum sensing comes from the following three parts. Firstly, in the traditional IRLS algorithm, the inverse of $\left(\boldsymbol{h}_f\boldsymbol{\Theta}\mathbf{W}^{(l)}\boldsymbol{\Theta}^T + \lambda_L\mathbf{I}_P\right)$ takes $\mathrm{O}\left(P^3\right)$ and it is required in each iteration. In large size CS problem, solving a problem with complexity $\mathrm{O}\left(P^3\right)\,I_{\max}$ times is unacceptable. As summarized in Algorithm 1, the proposed DNRLS-based CS algorithm solves the signal recovery problem in a non-iterative approach. Therefore, the computational complexity is $1/I_{\max}$ of the traditional IRLS-based compressive spectrum sensing in which $I_{\max}$ iterations are required to get an accurate estimation of the spectrum. Secondly, the computational complexity reduction is contributed by the fewer measurements required by the proposed DNRLS algorithm to achieve exact signal recovery. In the proposed DNRLS algorithm, the minimal number of measurements P for exact recovery

---

Algorithm 1 Data-driven non-iteratively reweighted least squares-based compressive spectrum sensing

---

Ensure: $p, \lambda, \boldsymbol{\Theta}, \mathbf{x}, \zeta, \bar{\boldsymbol{\gamma}}(t)$.
1: Calculate $P_{IB}(t+1)$ by the proposed Wilkinson's method-based DTT location probability model introduced in Sect. 3.2.3.
2: Map $P_{IB}(t+1)$ to $\boldsymbol{\gamma}(t+1)$.
3: Calculate $\bar{\boldsymbol{\gamma}}(t+1)$ by $\bar{\boldsymbol{\gamma}}(t)$ and $\boldsymbol{\gamma}(t+1)$ based on (3.7).
4: Perform signal recovery by (3.8) to get $\hat{\mathbf{s}}_{\mathbf{f}}$.
5: Make decision $\mathbf{d}$ on spectrum occupancy by comparing $\hat{\mathbf{s}}_{\mathbf{f}}$ with λ defined in (2.11).
6: **return d**.

---

is reduced to $\tilde{P}\left(\tilde{P} < P\right)$. It leads to a large computational complexity reduction as the complexity of solving the inverse of $\left(\boldsymbol{h}_f \boldsymbol{\Theta} \mathbf{W}^{(l)} \boldsymbol{\Theta}^T + \lambda_L \mathbf{I}_P\right)$ is $O\left(\tilde{P}^3\right)$. The performance analyses are further shown in numerical analyses. Thirdly, the computational complexity reduction comes from the calculation of P_{IB} in the proposed DNRLS framework. Specifically, to minimize the necessary computational complexity at SUs, the Wilkinson's method is utilized to calculate the P_{IB} for each TVWS channel. The details of the Wilkinson's method-based DTT location probability calculation algorithm are introduced in Sect. 3.2.3.

3.2.3 Proposed Wilkinson's Method-Based DTT Location Probability Calculation Algorithm

At an SU, the calculation of maximum allowable EIRP P_{IB} of each channel in TVWS should be efficient and accurate. Monte Carlo method and Schwartz-Yeh's method are the two algorithms approved by regulators to calculate the maximum allowable EIRP P_{IB}. Schwartz-Yeh's method is an approximate algorithm in which infinite loops are used to calculate the mean and standard deviation of lognormal distribution variables. However, the large computational complexity and low efficiency of the Schwartz-Yeh's method are difficult to overcome at power-limited SUs. In this chapter, the Wilkinson's method is invoked to calculate q_1, q_2, and P_{IB} in a much more efficient way.

3.2.3.1 Maximum Allowable Equivalent Isotropic Radiated Power Calculation

Based on the Wilkinson's method, q_1 and q_2 can be calculated accordingly. Taking the calculation of q_1 as an example, $\frac{P_{as,min}}{P_{as}} + \frac{V}{P_{as}} = A + B \leq 1$. $10\log_{10}(A+B) \leq 0$, which is equivalent to $X_{(dB)} = 10\log_{10}\left(10^{\frac{A_{dB}}{10}} + 10^{\frac{B_{dB}}{10}}\right) \leq 0$. It can be fitted into the precondition of Wilkinson's method to get $10^{\frac{A_{dB}}{10}} + 10^{\frac{B_{dB}}{10}} = 10^{X_{dB}} = e^{\Lambda_1} + e^{\Lambda_2}$. Therefore, $\Lambda_1 = \rho \times A_{(dB)}$ and $\Lambda_2 = \rho \times B_{(dB)}$. The relevant correlation coefficient of A and B can be given as

$$r_{A,B} = \frac{\mathrm{cov}\left(A_{(dB)}, B_{(dB)}\right)}{\sqrt{\mathrm{var}\left(A_{(dB)}\right)\mathrm{var}\left(B_{(dB)}\right)}} = \frac{\sigma_{as}}{\sqrt{\sigma_{as}^2 + \sigma_V^2}}, \tag{3.11}$$

where σ_{as} and σ_V can be calculated based on the DTT transmitter information used for geolocation database calculation algorithm.

Similarly, q_2 can be calculated by the Wilkinson's method by the following procedure:

1. Input $m_{as}, \sigma_{as}, m_V, \sigma_V, m_C$, and σ_C, which can be calculated based on the DTT transmitter information used for geolocation database calculation algorithm;
2. Calculate m_D and σ_D by Wilkinson's method based on m_V, σ_V, m_C, and σ_C;
3. Calculate m_A, σ_A, m_E, and σ_E by Wilkinson's method based on m_{as}, σ_{as}, m_D, and σ_D;
4. Calculate m_Y and σ_Y by Wilkinson's method based on m_A, σ_A, m_E, and σ_E;
5. Calculate q_2 based on m_A, σ_A, m_E, and σ_E.

With q_1 and q_2 calculated by the Wilkinson's method, the procedure of calculating P_{IB} is shown in Fig. 3.2. Firstly, input the mean and standard derivation of the received power of wanted DTT signal, i.e., P_{as}, and the minimum required power of wanted DTT signal, i.e., V, which can be obtained from the DTT transmitter information used for geolocation database calculation algorithm. As defined in IEEE 802.22 standard, the maximum allowable EIRP that can be utilized in TV frequency band is 4 W. Therefore, the predefined maximum allowable value (4 W) is assigned to P_{IB} for each TVWS channel. Subsequently, the mean and standard derivation

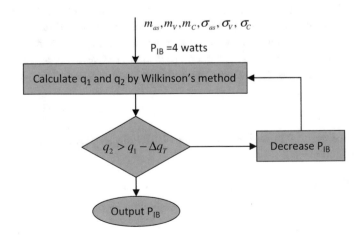

Fig. 3.2 The procedure of calculating maximum allowable P_{IB}

of C can be calculated based on initial value of P_{IB}. Additionally, q_1 and q_2 are calculated by the Wilkinson's method with inputs m_{as}, m_V, m_C, σ_{as}, σ_V, σ_C. Consequently, the corresponding P_{IB} is updated, which is utilized to calculate the new q_1 and q_2 until $q_2 \leq q_1 - \Delta q_T$. The output of this procedure is the maximum allowable EIRP P_{IB} for each TVWS channel.

3.3 Numerical Analyses

The analyses of the proposed stand-alone DNRLS framework on the simulated signals and data are presented in this section. Furthermore, the proposed framework is tested on the real-world signals collected by RFeye node and the data obtained from the geolocation database provided by Nominet.

3.3.1 Numerical Analyses on Simulated Signals and Data

In the simulations, OFDM signals are simulated as PUs, which is used by the DVB-T signals in TVWS from 470 to 790 MHz in the UK. There are a total of 40 channels in TVWS with a bandwidth of 8 MHz for each channel. It is assumed that each PU is independent and only locates at one channel. The transmission channel for signals is modeled as an AWGN channel. The target P_f is set to be 0.01.

The comparison of the proposed and traditional methods for calculating maximum allowable EIRP is presented firstly. Since Monte Carlo simulation is based on no assumption and approximation, its results can be considered precise as long as the number of trials is large enough. With 10,000 points, Monte Carlo simulation shows a relatively stable performance. By taking the results obtained by Monte Carlo simulation as a benchmark, the accuracy of the Schwartz-Yeh's method and Wilkinson's method can be measured by the error rate $\Delta Q(\cdot)/Q_{(MonteCarlo)}(\cdot)$, where $Q_{(MonteCarlo)}(\cdot)$ refers to values calculated by Monte Carlo simulation and $\Delta Q(\cdot)$ refers to the absolute difference of parameters' values between Schwartz-Yeh's method or Wilkinson's method and the Monte Carlo simulation. More specifically, $\Delta Q(q_1) = \left| q_1^{S,W} - q_1^M \right|$, $\Delta Q(q_2) = \left| q_2^{S,W} - q_2^M \right|$ and $\Delta Q(P_{IB}) = \left| P_{IB}^{S,W} - P_{IB}^M \right|$, where $q_1^{S,W}$, $q_2^{S,W}$, and $P_{IB}^{S,W}$ refer to the corresponding values calculated by the Schwartz-Yeh's method or Wilkinson's method, respectively, and q_1^M, q_2^M, and P_{IB}^M refer to the corresponding values calculated by Monte Carlo simulation. The error rates of q_1, q_2, and P_{IB} calculated by the Schwartz-Yeh's method and Wilkinson's method are shown in Table 3.1. It shows that the proposed Wilkinson's method outperforms the Schwartz-Yeh's method in terms of the calculation accuracy.

Similarly as the error rate calculation, running time of Monte Carlo simulation with 10,000 points is chosen as a benchmark when measuring the running time for

Table 3.1 Error rates comparison

	q_1	q_2	P_{IB}
Schwartz-Yeh's method	31.25%	4.76%	7.87%
Wilkinson's method	9.36%	1.31%	1.54%

Table 3.2 Running time comparison

	q_1	q_2	P_{IB}
Schwartz-Yeh's method	15,966.04%	153,278.65%	75,462.57%
Wilkinson's method	99.06%	98.89%	99.47%

Table 3.3 Comparison of actual maximum allowable EIRP P_{IB} in Oxford

	Actual maximum allowable EIRP P_{IB} (Watt)			
	The latest release of Ofcom TV white space model by Wilkinson's method			
Available channel	Open	Suburban	Urban	Power control model
22	0	4.0000	4.0000	4.0000
25	0	4.0000	4.0000	4.0000
28	0	4.0000	4.0000	4.0000
29	0.0025	4.0000	4.0000	4.0000
40	0	4.0000	4.0000	4.0000
43	0	4.0000	4.0000	4.0000
46	0	4.0000	4.0000	4.0000
49	0.0013	4.0000	4.0000	4.0000
51	0.3981	1.2589	4.0000	0.0002
54	0.0013	4.0000	4.0000	4.0000
58	0.0013	4.0000	4.0000	4.0000

the calculation of q_1, q_2, and q_{IB}. Table 3.2 shows the running time comparison of the Schwartz-Yeh's and Wilkinson's methods. It can be observed that the Wilkinson's method reduces the running time significantly in comparison with the Schwartz-Yeh's method. Therefore, the proposed Wilkinson's method is very suitable for SUs with limited power to obtain the q_1, q_2, and P_{IB} efficiently.

After validating the accuracy and efficiency, a national grid reference (NGR)-based geolocation database is built with the proposed Wilkinson's method. By utilizing the DTT transmitter information for geolocation database calculation algorithm, P_{IB} can be calculated by the proposed Wilkinson's method-based DTT location probability model for any specific location. Taking an NGR number of SP515065 in Oxford as a test location, the maximum allowable EIRP calculated by the power control and the proposed location probability model are shown in Table 3.3.

As shown in Table 3.3, there are 11 available channels at SP515065 in total. In the proposed location probability model, the transmission environment is classified into three situations: open, suburban, and urban. Coupling gain in different situations is treated differently, leading to different interference toleration levels of DTT

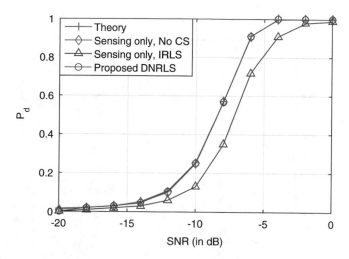

Fig. 3.3 Detection performance on the simulated signals and data under different SNR values, $p = 0.1$, compression ratio $\gamma = 20\%$

receivers. It is obvious that the power attenuation in open areas is much lower than suburban and urban areas. As a result, the actual maximum allowable EIRP P_{IB} in open areas is smaller than the other two situations at a certain NGR location. Taking channel 51 as an example, the P_{IB} is 0.0002 W in power control model. However, the spectrum of interest could be utilized more effectively if the transmission environment is classified, which is 0.3981 W in open areas, 1.2589 W in suburban areas, and 4.0000 W in urban areas.

Based on the obtained P_{IB} from the local geolocation database algorithm, the weights are constructed by fusing the current P_{IB} with historical data in the proposed DNRLS-based compressive spectrum sensing. Figure 3.3 shows detection performance of the sensing only approach and the proposed DNRLS framework implemented at SUs, where p is set to be 0.1. It is observed that the detection performance of the sensing only approach without CS implemented at an SU is matched with the theoretical curve, which is presented as a benchmark and expressed as (4.6).

Figure 3.3 shows that detection performance of the sensing only approach with IRLS is smaller than the theoretic curve due to the signal recovery errors caused by the sub-Nyquist sampling ($\gamma = 20\%$). When the proposed DNRLS framework is performed, detection probability increases greatly which can almost match with the theoretic curve. The reason for the large performance improvement is that the data used to construct the weights is the exact representation of the spectrum of interest if there is no unregistered user. Additionally, it is noted that the sensing only approach with IRLS requires an iterative process to update the weights. This iterative process introduces a higher computational complexity. As a result, the proposed DNRLS-based compressive spectrum sensing can achieve better detection performance with $(I_{max} - 1)/I_{max} - 1$ of computational complexity reduced in comparison with the sensing only approach with IRLS.

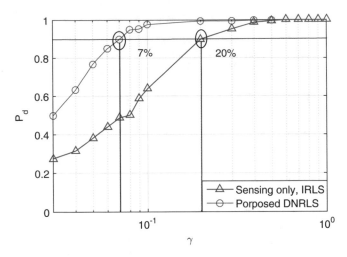

Fig. 3.4 Detection performance on the simulated signals and data under different compression ratios γ, $p = 0.1$, SNR $= -5$ dB

Figure 3.4 shows detection probability of the sensing only approach with IRLS and the proposed DNRLS framework with varying compression ratios. In this scenario, the spectrum occupancy ratio is assumed to be 12.5%, p is 0.1, and the SNR value is -5 dB. It is noted that there is a big difference on the necessary number of measurements between the proposed DNRLS framework and the sensing only approach to achieve the same detection probability. Specifically, as shown in Fig. 3.4, the proposed DNRLS framework can achieve 90% detection probability when the compression ratio is no higher than 7%. However, the sensing only approach requires the compression ratio to be about 20% in order to achieve the same performance. As a result, the sampling rates can be reduced by 13% without degrading the detection performance.

The detection performance of the proposed DNRLS framework is shown in Fig. 3.5 with different spectrum occupancy ratios in TVWS and different p values for l_p. In this scenario, SNR is set to be -5 dB and the positions of these active PUs are set to be random. In compressive spectrum sensing, increasing spectrum occupancy in spectrum of interest refers to higher sparsity levels of the signal to be recovered. It can be observed that the detection performance becomes improved with decreasing value of p and fixed sparsity level. Meanwhile, the detection performance is degraded slightly with increasing sparsity level increases when the value for p is fixed. As a result, more compressed measurements should be collected at SUs to avoid performance degradation when sparsity level increases.

Figure 3.6 shows the detection probability of the proposed DNRLS framework under different window sizes T with new unregistered users showing up in the spectrum of interest. In this scenario, the spectrum occupancy is 12.5%, p is 0.1, and compression ratio is 10%. With unregistered users in TVWS, only half of the weights for the channels with active PUs are exact. It can be observed that the

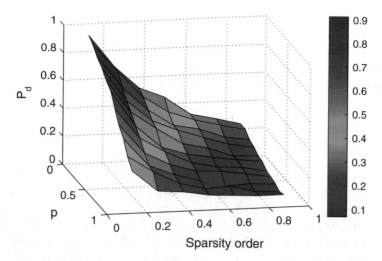

Fig. 3.5 Detection performance on the simulated signals and data under different sparsity levels and p values, compression ratio $\gamma = 10\%$, SNR $= -5$ dB

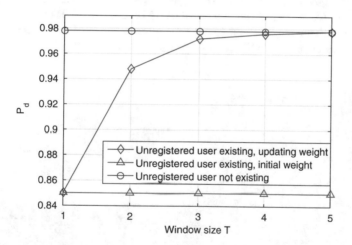

Fig. 3.6 Detection performance on the simulated signals and data under different window sizes T with unregistered users existing, compression ratio $\gamma = 10\%$, $p = 0.1$, SNR $= -5$ dB

detection performance is degraded from 98 to 85% in the first sensing period after a new unregistered user shows up in TVWS. However, after one sensing period has passed, which refers to $T = 2$, the detection performance is improved to about 95%. This improvement benefits from the weights are constructed by fusing the output of the geolocation database algorithm with the historical data. The geolocation database algorithm utilizes the self-maintained geolocation database at SU locally which contains the new unregistered users' information. Furthermore, the detection performance converges to 98% after four updates of the weights. With increasing window size T, the improvement on detection performance becomes slower after the first updating on the weights. However, if the unregistered user shows up again in the same position of TVWS, detection probability of the proposed DNRLS framework falls between 85 and 95%, which is dependent on the window size T. If T is large enough, the detection probability would get close to 95%.

3.3.2   Numerical Analyses on Real-World Signals and Data

After the proposed robust single node spectrum sensing algorithm has been validated with simulated signals, it is further tested on the real-world signals recorded by the RFeye node and the real data provided by the geolocation database from Nominet qualified by Ofcom. The RFeye node is a scalable and cost-effective node which can provide real-time 24/7 monitoring of radio spectrum. It is capable of sweeping spectrum from 10 to 6 GHz, and can capture signals of all types, including transient transmission such as pulsing or short-burst signals. It is even sensitive to very low power signals. The RFeye node used for measurement is located at Queen Mary University of London (QMUL) (51.523021°N 0.041592°W) as shown in Fig. 3.7 with the height about 15 m above ground. The real-world signal recorded by the RFeye node is for TVWS ranging from 470 to 790 MHz.

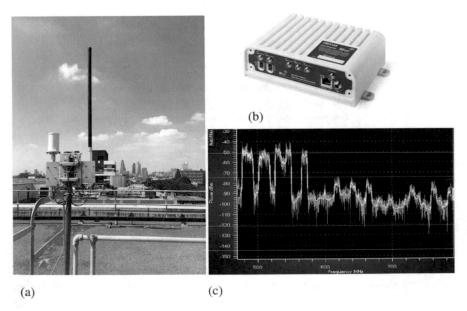

Fig. 3.7 Measurement setup for real-world TVWS signals recorded at Queen Mary University of London. (**a**) Measurement setup at Queen Mary University of London. (**b**) RFeye sensing node. (**c**) Real-time TVWS signal observed by RFeye sensing node

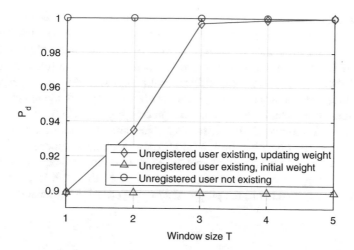

Fig. 3.8 Detection performance on the real-world signals under different window sizes T with unregistered users existing, compression ratio $\gamma = 10\%$, $p = 0.1$

Some pilots over TVWS have been undertaken in the UK as launched by the Ofcom. In the trials run at QMUL, an unregistered user is transmitted in TVWS channel 27 (518 to 526 MHz). In this case, the historical data and P_{IB} from the geolocation database would not be exact for the channel 27 as it is the first time for the unregistered user showing up in TVWS. As a result, the output of the geolocation database would still allow a high EIRP in channel 27. The simulation results for the case with unregistered users under different window sizes T are shown in Fig. 3.8. It can be observed that the detection performance would be degraded once the unregistered user shows up in TVWS. This is caused by the inexact weights constructed by the inaccurate P_{IB} in channel 27. Similarly as Fig. 3.6, the detection performance is increased largely after window size T is increased to 2. With increasing window size, the detection performance of the proposed DNRLS framework converges efficiently.

Based on the fast convergence performance shown in Figs. 3.6 and 3.8, it can be indicated the practicability of the proposed DNRLS framework is reasonable. The implementation of compressive spectrum sensing with a geolocation database algorithm can improve the energy efficiency at SUs by reducing its computational complexities. Therefore, such an energy efficient algorithm could be applied to multiple scenarios with energy-constrained devices.

3.4 Summary

This chapter introduced a stand-alone DNRLS framework combining compressive spectrum sensing with geolocation database for wideband spectrum. In particular, a DNRL-based compressive spectrum sensing algorithm was proposed to reduce the sampling rates and lower the computational complexities by invoking geolocation database. Additionally, the proposed framework was tested on the real-world signals and data after having been validated by the simulated signals and data over TVWS. The numerical results showed that the computational complexities of signal recovery process were reduced with improved detection performance. Furthermore, it is noted the proposed framework can also provide benefits to relax the requirement on sparsity level estimation in compressive spectrum sensing. More recently, an autonomous CS-based sensing algorithm has been proposed in Zhang et al. (2018), which enables the local SU to automatically choose the minimum sensing time without knowledge of spectral sparsity or channel characteristics. The compressive samples are collected block-by-block in time while the spectral is gradually reconstructed until the proposed stopping criterion is reached.

References

Ba, D., Babadi, B., Purdon, P., & Brown, E. (2014). Convergence and stability of iteratively reweighted least squares algorithms. *IEEE Transactions on Signal Processing, 62*, 183–195.

Candes, E. (2006). Compressive sampling. In *Proceedings of the International Congress of Mathematicians*, Madrid, Spain (Vol. 3, pp. 1433–1452).

Candes, E. J., Wakin, M. B., & Boyd, S. P. (2008). Enhancing sparsity by reweighted l_1 minimization. *Journal of Fourier Analysis and Applications, 14*, 877–905.

Carrillo, R. E., & Barner, K. (2009). Iteratively re-weighted least squares for sparse signal reconstruction from noisy measurements. In *Proceedings of Conference on Information Sciences and Systems (CISS)*, Baltimore, MD (pp. 448–453).

Chartrand, R. (2007). Exact reconstruction of sparse signals via nonconvex minimization. *IEEE Signal Processing Letters, 14*, 707–710.

Chartrand, R., & Staneva, V. (2008). Restricted isometry properties and nonconvex compressive sensing. *Inverse Problems, 24*, 035020.

Chartrand, R., & Yin, W. (2008). Iteratively reweighted algorithms for compressive sensing. In *Proceedings of IEEE International Conference on Acoustics, Speech, and Signal Processing (ICASSP)*, Las Vegas, NV (pp. 3869–3872).

Escoda, O., Granai, L., & Vandergheynst, P. (2006). On the use of a priori information for sparse signal approximations. *IEEE Transactions on Signal Processing, 54*, 3468–3482.

Fenton, L. (1960). The sum of log-normal probability distributions in scatter transmission systems. *IRE Transactions on Communications Systems, 8*, 57–67.

Friedlander, M., Mansour, H., Saab, R., & Yilmaz, O. (2012). Recovering compressively sampled signals using partial support information. *IEEE Transactions on Information Theory, 58*, 1122–1134.

Holland, O., Ping, S., Aijaz, A., Chareau, J., Chawdhry, P., Gao, Y., et al. (September 2015). To white space or not to white space: That is the trial within the ofcom TV white spaces pilot. In *Proceedings of the IEEE International Symposium Dynamic Spectrum Access Networks (DYSPAN)*, Stockholm, Sweden (pp. 11–22).

Kolodzy, P., & Avoidance, I. (June 2002). Spectrum policy task force. In *Report ET Docket*. Washington, DC:Federal Communications Commission.

Lu, W., & Vaswani, N. (2012). Regularized modified BPDN for noisy sparse reconstruction with partial erroneous support and signal value knowledge. *IEEE Transactions on Signal Processing, 60*, 182–196.

Miosso, C. J., von Borries, R., Argaez, M., Velázquez, L., Quintero, C., & Potes, C. (2009). Compressive sensing reconstruction with prior information by iteratively reweighted least-squares. *IEEE Transactions on Signal Processing, 57*, 2424–2431.

Paisana, F., Marchetti, N., & DaSilva, L. (2014). Radar, TV and cellular bands: Which spectrum access techniques for which bands? *IEEE Communications Surveys & Tutorials, 16*, 1193–1220.

Ribeiro, J., Ribeiro, J., Rodriguez, J., Dionisio, R., Esteves, H., Duarte, P., et al. (2012). Testbed for combination of local sensing with geolocation database in real environments. *IEEE Wireless Communications, 19*, 59–66.

RFeye node https://www.crfs.com/all-products/hardware/nodes/.

Saab, R., & Yılmaz, Ö. (2010). Sparse recovery by non-convex optimization–instance optimality. *Applied and Computational Harmonic Analysis, 29*, 30–48.

Sun, H., Chiu, W.-Y., & Nallanathan, A. (2012). Adaptive compressive spectrum sensing for wideband cognitive radios. *IEEE Communications Letters, 16*, 1812–1815.

Tian, Z., & Giannakis, G. (2007). Compressed sensing for wideband cognitive radios. In *Proceedings IEEE International Conference on Acoustics, Speech, and Signal Processing (ICASSP)*, Honolulu, HI (pp. 1357–1360).

Wang, Y., Tian, Z., & Feng, C. (2012). Sparsity order estimation and its application in compressive spectrum sensing for cognitive radios. *IEEE Transactions on Wireless Communications, 11*, 2116–2125.

Wang, J., Ding, G., Wu, Q., Shen, L., & Song, F. (2014). Spatial-temporal spectrum hole discovery: A hybrid spectrum sensing and geolocation database framework. *Chinese Science Bulletin, 59*, 1896–1902.

Wang, N., Gao, Y., & Evans, B. (2015). Database-augmented spectrum sensing algorithm for cognitive radio. In *Proceedings of IEEE International Conference on Communications (ICC)*, London, UK (pp. 7468–7473).

Wipf, D., & Nagarajan, S. (2010). Iterative reweighted l_1 and l_2 methods for finding sparse solutions. *IEEE Journal of Selected Topics in Signal Processing, 4*, 317–329.

Zhang, X., Ma, Y., Gao, Y., & Zhang, W. (2018). Autonomous compressive sensing augmented spectrum sensing. *IEEE Transactions on Vehicular Technology, 67*, 6970–6980.

Chapter 4
Robust Compressive Spectrum Sensing

In this chapter, the existing work on compressive spectrum sensing in CRNs and the main contributions are firstly reviewed in Sect. 4.1. In Sect. 4.2, the proposed robust compressive spectrum sensing working at a single CR user is presented. Section 4.3 gives the related simulation results. Additionally, the proposed robust sub-Nyquist spectrum sensing algorithm for the CSS scenario is demonstrated in Sect. 4.4, in which the low-rank MC technique is invoked to perform signal recovery. The numerical results are presented in Sect. 4.5. Finally, Sect. 4.6 concludes this chapter.

4.1 Introduction

4.1.1 Related Work

It has been identified that the CS-enabled system is somewhat sensitive to noise, exhibiting a 3 dB SNR loss per octave of subsampling (Treichler et al. 2009), which parallels the classic noise-folding phenomenon and makes exact signal recovery more difficult for the case with high channel noise. Therefore, a robust compressive spectrum sensing algorithm with low computational complexity is more than desired.

Moreover, single node spectrum sensing faces the challenges that detection performance is significantly degraded if an SU experiences multipath fading and hidden terminals (Digham et al. 2007; Akyildiz et al. 2011). This may cause miss detection. As a result, the SU may unwittingly transmit signals in channels with active PUs, which may cause serious interference to the PUs. In order to reduce the influence of imperfect channel environment, multiple nodes spectrum sensing, named as CSS, was proposed to efficiently combat fading problems by utilizing a

Y. Gao, Z. Qin, *Data-Driven Wireless Networks*, SpringerBriefs in Electrical and Computer Engineering, https://doi.org/10.1007/978-3-030-00290-9_4

spatial diversity of cooperative multiple SUs (Quan et al. 2009; Zeng et al. 2011; Ghasemi and Sousa 2005).

In CSS networks, there are two types of data fusion: centralized and decentralized fusion. In decentralized CSS, each SU only communicates with its neighbor SUs within one hop to reduce the transmission power consumed during sensing. After convergence, all SUs will have the fused sensing result without the implementation of an FC. Several decentralized CSS schemes (Tian 2008; Bazerque and Giannakis 2010; Li et al. 2010) have been proposed where the average value of all the local spectrum sensing decisions is computed to get the final decision. As a result, the final decision obtained might be sub-optimal. Additionally, Zeng et al. (2011) proposed a distributed CSS algorithm in which sensing samples rather than sensing decisions are exchanged with the neighbor SUs within multi-hops to reach a global fusion at the cost of increasing network load. The convergence speed of CSS is an issue in large-scale networks. Moreover, Zhang et al. (2018) proposed a blind joint sub-Nyquist sensing scheme by utilizing the surround IoT devices to jointly sample the spectrum based on the multi-coset sampling theory. Thus, only the off-the-shelf low-rate ADCs on the IoT devices are required to form coset samplers and only the minimum number of coset samplers are adopted without the prior knowledge of the number of occupied channels and signal-to-noise ratios.

In the centralized CSS scheme, all SUs report to an FC to make a final decision. In Ma et al. (2016), an SU senses the whole spectrum of interest, and then the SU sends all the collected compressed measurements to an FC to get a global decision. As a result, the optimal decision can be obtained but the transmission load in the reporting channel between SUs and the FC is heavy. In Wang et al. (2012), in order to reduce the sampling costs and transmission load between SUs and the FC, the length of received signal's frequency domain representations is set to be equal to the number of channels in the spectrum of interest rather than the original length of received signal in time domain, which results in a very poor resolution in the frequency domain and serious spectral leakage in each channel. Consequently, the P_f would increase and the P_d would decrease. Additionally, as aforementioned, the noise becomes critical after signals are collected at sub-Nyquist rate (Treichler et al. 2009). Therefore, a robust sub-Nyquist sampling-based CSS algorithm with high spectrum resolution and low computational complexity is required.

4.1.2   Contributions

The main contributions of this chapter are summarized as follows:

1. A robust compressive spectrum sensing algorithm is proposed, in which the data acquisition and signal recovery are both conducted at an SU locally. In the proposed algorithm, the computational complexity is significant reduced by a new channel division scheme. Additionally, a denoising algorithm is performed to improve detection performance and make the algorithm robust to channel noise.

2. A robust sub-Nyquist sampling-based CSS algorithm is proposed to reduce the signal acquisition costs at SUs, improve the spectrum resolution and the robustness to channel noise, by invoking the low-rank MC technique. In the considered system, signal recovery is performed at an FC. In the proposed algorithm, the computational complexity is reduced significantly by the new channel division scheme. Additionally, the robustness to channel noise is improved by the proposed denoising algorithm.
3. The proposed robust sub-Nyquist sampling-based spectrum sensing algorithms are both tested on the real-world signals over TVWS after being validated by the simulated TV signals.

4.2 Robust Compressive Spectrum Sensing at Single User

4.2.1 System Model

In the single node case, the compressive spectrum sensing model is same as aforementioned in Sect. 2.4.3 of Chap. 3. As no prior information of PUs is required, l_1 norm minimization is invoked to perform signal recovery. In order to reduce the computational complexity during signal recovery process and enhance algorithm's robustness to imperfect channel noise, a robust spectrum sensing algorithm is proposed for the single node case based on CS. In the first phase, a new efficient channel division scheme is proposed to reduce the computation complexity for signal recovery. In the second phase, a denoising algorithm is proposed to make the algorithm robust against high channel noise.

4.2.1.1 Proposed Channel Division Scheme

When an l_1 norm minimization-based spectrum sensing algorithm is implemented at an SU, the computational complexity of signal recovery is dependent on the number of samples to be recovered. In the considered model, it is assumed that the spectrum of interest can be divided into $\mathscr{I}$ channels. A new channel division scheme is proposed, in which only L ($L < \mathscr{I}$) out of $\mathscr{I}$ channels are expected to be sensed in one sensing period at SUs to reduce the number of samples to be recovered. As shown in Fig. 4.1, each L-channel group is indexed by i $\left(i = 1, 2, \cdots, \frac{\mathscr{I}}{L}\right)$. If any vacant channel is detected, SU would stop sensing and start data transmission. Otherwise, SU begins to sense the next L-channel group in the next sensing period. As a result, the required sampling rates at SUs for exact recovery are further reduced by implementing the CS technique at SUs.

Once signal of the L-channel group $\mathbf{r_{fi}} = \mathbf{h_{fi}s_{fi}} + \mathbf{w_{fi}} \in \mathscr{C}^{n \times 1}$ ($n = \frac{LN}{\mathscr{I}}$) arrives at the receiver, where N is the number of samples for the whole spectrum at Nyquist rates. The compressed measurements $\mathbf{x_i} \in \mathscr{C}^{\tilde{p} \times 1}$ are collected at sub-Nyquist

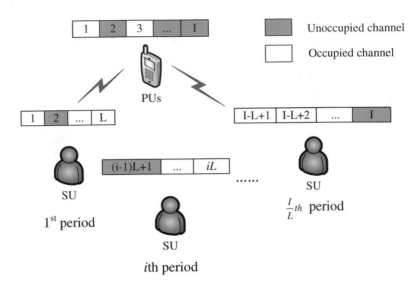

Fig. 4.1 System model of the proposed channel division scheme in the single node spectrum sensing based on compressive sensing

sampling rates, with compression ratio is defined as $\gamma = \frac{\tilde{p}}{n}$. Here, $\mathbf{h_{fi}}$, $\mathbf{s_{fi}}$, $\mathbf{w_{fi}}$ refer to the frequency domain representations of channel coefficients, transmitted primary signal, and channel noise in the L-channel group channel. The noise variance is σ_n^2. Subsequently, the reconstructed frequency domain representations of the ith L-channel group $\hat{\mathbf{s}}_{\mathbf{fi}}$ can be obtained by solving l_1 norm minimization as:

$$\begin{aligned} &\min \; \left\| \hat{\mathbf{s}}_{\mathbf{fi}} \right\|_1, \\ &\text{subject to} \; \left\| \boldsymbol{\Theta}_i \cdot \mathbf{h_{fi}} \hat{\mathbf{s}}_{\mathbf{fi}} - \mathbf{x_i} \right\|_2^2 \leq \varepsilon, \end{aligned} \quad (4.1)$$

where $\boldsymbol{\Theta}_i \in \mathscr{C}^{\tilde{p} \times n}$, and ε is the tolerance for noise level.

4.2.1.2 Proposed Denoised Spectrum Sensing Algorithm

When making a decision for spectrum occupancy, the decision accuracy is influenced by the signal recovery errors. The recovery performance of traditional l_1 norm minimization algorithm is degraded by high channel noise and low compression ratio. Furthermore, it is noticed that the amplitudes of recovered signal $\hat{\mathbf{s}}_{\mathbf{fi}}$ may be negative with high absolute values. Here, the nonpositive power spectrum amplitudes are caused by the unsuccessful signal recovery after solving problem (4.1). The first possible reason for unsuccessful signal recovery is that the number of collected measurements $\tilde{p}$ is not enough for exact signal recovery. The second possible reason for unsuccessful signal recovery is caused by the high noise level that dominates the compressed measurements $\mathbf{x_i}$. However, the power spectrum $\mathbf{s_{fi}}$

is nonnegative. If those negative values are used to calculate the energy density, it would become higher than the real energy value. As a result, the P_f of spectrum sensing would increase, which means the vacant channels might be determined as occupied. In order to improve the recovery performance and detection performance, a denoising algorithm is proposed.

In the denoising algorithm, the amplitude of the bth sample in the recovered signal $\hat{s}_{fi}$ is compared with the corresponding noise level $\sigma_n(b)$, where b $(1 \leq b \leq n)$ is the index of the recovered signal. If $\hat{s}_{fi}(b)$ is higher than $\sigma_n(b)$, the compressed measurement collected at SUs $r_{fi}(b)$ is kept for the recovered signal. Otherwise, the corresponding value will be set to zero to reduce the recovery error. Here, the recovery error may be caused by the high channel noise or not enough number of compressed measurements. The denoised signal $\hat{s}_{fi\_d}$ can be expressed as:

$$\hat{s}_{fi\_d}(b) = \begin{cases} r_{fi}(b), & \text{if } \hat{s}_{fi}(b) \geq \sigma_n(b), \\ 0, & \text{otherwise.} \end{cases} \tag{4.2}$$

After the denoising algorithm is performed, the energy density of each considered L-channel group in the denoised signal is compared with the corresponding threshold as defined in (2.11) to determine the spectrum occupancy of the corresponding L-channel group. If any L-channel group are determined as vacant, they can be used by SUs to transmit the unlicensed signals. Otherwise, the SU should continue sensing the next L-channel group until any vacant channel is found out or the $\frac{\mathscr{I}}{L}$ sensing periods, named as a sensing loop, are run out. As there is a high probability that the spectrum vacant in last loop remains free in the current sensing loop, an SU should firstly sense the L-channel group determined as free in the last sensing loop at the beginning of a new sensing loop if any vacant L-channel group are detected in the most recent sensing loop. Otherwise, an SU should keep sensing from the first L-channel group.

4.2.2 Computational Complexity and Spectrum Usage Analyses

In compressive spectrum sensing algorithm, the computational complexity mainly comes from the signal recovery process by solving the l_1 norm minimization problem. It is determined by the number of samples (N) to be recovered to represent the spectrum of interest. Specially, when the whole wideband spectrum of interest is sensed in one sensing period by an SU, the computational complexity of solving the l_1 norm minimization problem can be expressed as:

$$C_1 = O\left(N^3\right). \tag{4.3}$$

In the adaptive compressive spectrum sensing algorithm (Sun et al. 2012) for wideband CRs, the required computational complexity C_2 can be expressed as follows. In order to simplify the comparison, the spectrum sensed in each sensing period is assumed to be L out of $\mathscr{I}$ channels and the system starts data transmission after i sensing periods.

$$
\begin{aligned}
C_2 &= \mathrm{O}\left(\left(\tfrac{L}{\mathscr{I}} \times N\right)^3 + \left(\tfrac{2L}{\mathscr{I}} \times N\right)^3 + \ldots + \left(\tfrac{iL}{\mathscr{I}} \times N\right)^3 \right) \\
&= \mathrm{O}\left(\left(1 + 2^3 + \ldots + i^3\right) \times \left(\tfrac{L}{\mathscr{I}}\right)^3 \times N^3 \right) \\
&= \mathrm{O}\left(\left(\tfrac{(1+i)i}{2}\right)^2 \times \left(\tfrac{L}{\mathscr{I}}\right)^3 \times N^3 \right),
\end{aligned}
\tag{4.4}
$$

where $i = 1, \ldots, \frac{\mathscr{I}}{L}$ is the number of sensing periods that an SU needs to perform exact signal recovery to determine the accessible channels.

When the proposed new channel division scheme is used for single node wideband spectrum sensing, computational complexity of the signal recovery process is expressed as:

$$
C_3 = \mathrm{O}\left(i \times \left(\frac{L}{\mathscr{I}}\right)^3 \times N^3 \right).
\tag{4.5}
$$

As analyzed above, an SU may need multiple sensing periods to find out the accessible spectrum holes. Assuming there is at least one vacant channel in the spectrum of interest, the required sensing periods by the proposed channel division scheme is dependent on the number of channels in an L-channel group and the number of active PUs among the spectrum of interest. The worst case for the proposed scheme is that an SU does not find any vacant channel until the $\frac{\mathscr{I}}{L}$th sensing period. In such a case, $C_3 = \mathrm{O}\left(\left(\frac{L}{\mathscr{I}}\right)^2 \times N^3 \right)$ as $i = \frac{\mathscr{I}}{L}$. In practice, there are multiple vacant channels in the spectrum of interest due to the low spectrum utilization. Therefore, the required sensing periods would be less than $\frac{\mathscr{I}}{L}$ in reality. As a result, $C_1 > C_3$ in all cases. The proposed channel division scheme relaxes the requirement on high-speed ADC at the expense of compromised spectrum usage efficiency. This tradeoff is shown in the simulation part in Sect. 4.3. It seems the tradeoff is acceptable as the proposed algorithm is designed for networks in which SUs have limited computational power and infrequent low-speed transmission requirements. Comparing C_2 and C_3, it shows $C_2 = C_3$ if $i = 1$, which refers to the scenario that vacant channels can be found after signal recovery is only performed once. Otherwise, $C_2 > C_3$. Therefore, the proposed channel division scheme achieves a lower computational complexity than existing algorithms.

4.3 Numerical Analyses for Single User Case

4.3.1 Analyses on Simulated Signals

In the simulation, signals are orthogonal frequency division multiplexed (OFDM) generated as PUs, which are used in digital video broadcasting-terrestrial (DVB-T) over TVWS spectrum from 470 to 790 MHz in the UK. There are $\mathscr{I} = 40$ channels in TVWS with bandwidth of 8 MHz for each channel. P_f is set to be 0.01. $SNR = \sigma_s^2/\sigma_n^2$ is the ratio of signal power over noise power of an L-channel group. In the following simulations, the aforementioned tradeoff between spectrum usage efficiency and computational complexity is demonstrated firstly. Additionally, the influence of compression ratio, sparsity order, and the classic receiver operating characteristics (ROC) curves are presented to validate the proposed algorithm.

Figure 4.2 shows the average number of sensing periods which is required at SUs to find out the vacant channel for unlicensed usage. As aforementioned, the size of L-channel group which is sensed in each sensing period at SUs would influence the spectrum usage efficiency of the proposed channel division scheme. If $L = 1$, the case becomes a narrow band spectrum sensing which requires low-speed sampling rates at SUs. But the spectrum usage efficiency is low. With increasing L, it becomes a multichannel wideband spectrums sensing case in which the spectrum usage efficiency is increased with cost of expensive sampling acquisition. From Fig. 4.2, it can be observed that the vacant channels can be detected efficiently even with increasing L. Here, sparsity level refers to the ratio of occupied channels and the total number of channels. Higher sparsity level refers to higher spectrum occupancy, as the active PUs would result in non-zero amplitude in frequency domain. With higher sparsity level, the average required sensing periods to find the

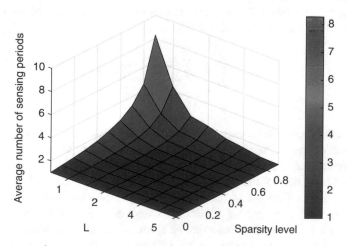

Fig. 4.2 Average number of required sensing periods at SUs with different sparsity levels and number of channels L sensed in one sensing period, SNR $= -5$ dB

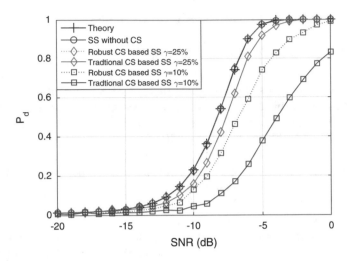

Fig. 4.3 Proposed robust compressive spectrum sensing algorithm at a single user achieves higher P_d than the traditional algorithms with simulated signals under different compression ratios γ and different SNR values

vacant spectrum holes increases. As the spectrum is underutilized in practice, the required number of sensing periods is relatively low. In the following simulation, it is assumed that the number of channels sensed by the SU in each sensing period is set to be $L = 8$.

Figure 4.3 shows P_d for the traditional l_1 norm minimization-based spectrum sensing algorithm, and the proposed robust single node spectrum sensing algorithm based on CS under different number of compressed measurements with varying SNR values. Its detection performance is also compared with that of spectrum sensing algorithm without CS implemented, as well as the theoretical values derived from Ye et al. (2008); Wang et al. (2013):

$$
P_d = Q\left(\frac{\frac{\lambda}{\sigma_n^2} - \left(1 + \frac{\sigma_s^2}{\sigma_n^2}\right)}{\left(1 + \frac{\sigma_s^2}{\sigma_n^2}\right)/\sqrt{n/2}} \right),
\tag{4.6}
$$

where λ is the threshold for energy detection as calculated by (2.11), and σ_s^2 refers to the power of transmitted primary signal. Here, n refers to the number of samples sampled from an L-channel group at Nyquist rate.

Figure 4.3 shows that the performance of l_1 norm minimization-based spectrum sensing algorithm (labeled as traditional CS-based SS) and the proposed robust single node spectrum sensing algorithm based on CS (labeled as robust CS-based SS) are both the same with that of spectrum sensing algorithm without CS implemented at the SU (labeled as no CS) and the theoretical curves obtained by (4.6). In Fig. 4.3, SS is the abbreviation for spectrum sensing which is only used

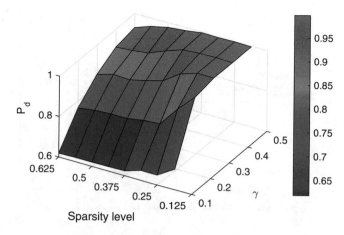

Fig. 4.4 P_d comparison with different sparsity levels and different compression ratios γ, SNR $= -5$ dB

in the legend. In this case, the number of occupied channels is 1 among 8. Therefore, the sparsity level is set to be 12.5%. When the number of collected measurements decreases, the detection performance degrades. It also shows that performance of the proposed robust single node spectrum sensing based on CS is better than that of the CS-based spectrum sensing without denoising when the compression ratio is 25% and 10%. This gain benefits from the proposed denoising algorithm which can improve the signal recovery accuracy. As the recovery accuracy becomes higher with the higher compression ratio, detection performance of the proposed robust spectrum sensing algorithm gets closer to the theoretical curves. The simulation result shows that the proposed robust spectrum sensing algorithm can reduce the sampling rates by 75% without degrading detection performance.

Figure 4.4 shows the detection performance of the proposed robust single node spectrum sensing based on CS with different sparsity levels and different compression ratios. In this scenario, the different sparsity levels refer to different number of active PUs in the spectrum of interest. The positions of these active PUs are set to be random. The detection performance becomes worse with increasing sparsity level and decreasing compression ratio as shown in Fig. 4.4. As the sparsity level increases, sparse property of signal to be recovered becomes less sparse, and therefore more compressed measurements should be collected for signal recovery to make sure the detection performance not being degraded. It is noticed that the detection performance would only be slightly degraded when the proposed algorithm is applied to the practical signals in TVWS spectrum as its occupancy ratio is normally 15 to 20% in practice Kolodzy and Avoidance (2002); Ofc (a).

The ROC curves under different SNR values are shown in Fig. 4.5, where the compression ratio is set to be 25%. In this case, the sparsity level is set to be 12.5%. It can be observed that the proposed robust spectrum sensing algorithm based on CS exhibits better performance than the traditional spectrum sensing algorithm based on CS. Meanwhile, it is also noticed that the performance of the proposed robust

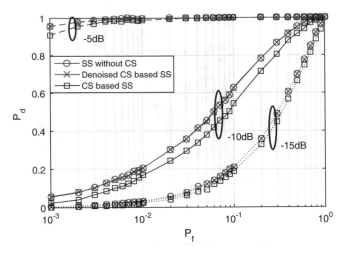

Fig. 4.5 Proposed robust compressive spectrum sensing algorithm at a single node achieves higher ROC curves than the traditional algorithms with simulated signals, and compression ratio $\gamma = 25\%$

spectrum sensing algorithm is almost as good as that of spectrum sensing algorithm without CS applied. This gain arises from the proposed denoising algorithm. This result matches with Fig. 4.3 when compression ratio is set to be 25%. It should be pointed that the increasing P_f refers to decreasing threshold level if the number of samples is fixed as defined in (2.11). Therefore, the detection performance becomes degraded with increasing threshold level as shown in Fig. 4.5.

4.3.2 Analyses on Real-World Signals

After the proposed robust single node spectrum sensing algorithm has been validated with simulated signals, it is further tested on the real-world signals recorded by the RFeye node as shown in Fig. 3.7 in Chap. 3.

When the recorded real-world signal is used as source signal for the proposed robust single node spectrum sensing algorithm, Fig. 4.6 shows P_d and P_f of the spectrum sensing without CS implemented, traditional CS-based spectrum sensing, and the proposed robust spectrum sensing algorithms under different threshold values. Here, the thresholds are experimental values. In this scenario, the compression ratio is set to be 15%. It can be observed that both P_d and P_f decrease with increasing threshold values. IEEE 802.22 demands a stringent sensing requirement. For the maximum P_f of 10%, a sensing algorithm should achieve 90% for P_d. According to Fig. 4.6, it shows that the detection performance of the spectrum sensing without CS implemented can achieve the target performance required in IEEE 802.22 when threshold is set to be -73.5 dBm or higher. However, the P_d of the algorithms with CS would be degraded with increasing threshold.

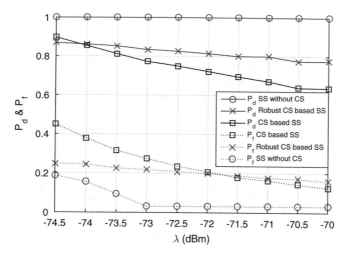

Fig. 4.6 P_d and P_f comparison of single node spectrum sensing with real-world signals under different thresholds λ, and compression ratio $\gamma = 15\%$

Therefore, -73.5 dBm is chosen as the suitable threshold to get a better tradeoff of P_d and P_f in the following analyses. From Fig. 4.6, it is also noticed that the proposed robust single node spectrum sensing algorithm outperforms the traditional one when threshold is 1.5. It can be observed that the P_d increases with decreasing threshold level which is matched with the simulation results shown in Fig. 4.5.

Figure 4.7 shows the P_d and P_f of the traditional spectrum sensing algorithm based on CS and the proposed robust spectrum sensing algorithm with real-world signals under different compression ratios from 1 to 100%. In this scenario, the threshold value is set to be -73.5 dBm according to Fig. 4.6. It can be observed that the detection performance gets better with increasing number of compressed measurements collected at the SU, and the proposed robust spectrum sensing algorithm outperforms the traditional one, which is similar with the results of simulated signals as shown in Fig. 4.3. It is further noticed that there is sharp dropping for P_f when the compression ratio γ is increased from 20 to 30%. This is caused by that the signal recovery becomes exact when the compression ratio γ is no less than 20%. Once the signal recovery is exact, P_f caused by recovery errors would be alleviated.

4.4   Matrix Completion-Based Robust Spectrum Sensing at Cooperative Multiple Users

Based on the proposed robust spectrum sensing algorithm for single node in CRNs, a new robust CSS algorithm based on low-rank MC is proposed to overcome the deep fading problem. In the considered network, the whole spectrum of interest

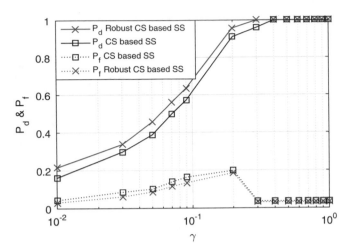

Fig. 4.7 Proposed robust compressive spectrum sensing algorithm at single node achieves higher P_d and lower P_f than the traditional algorithm with real-world signals under different compression ratios γ, and threshold λ is -73.5 dBm

can be divided into $\mathscr{I}$ channels, and K out of the I channels are occupied by PUs. It is assumed that the positions of active PUs are random among the whole spectrum of interest. The proposed algorithm includes two phases. In the first phase, the proposed channel division scheme is extended to the CSS scenario, in which each SU is implemented to sense an L-channel group of the $\mathscr{I}$ channels to reduce sampling rates. Here, each L channels are sensed by the same SU. As a result, at least $S(S = \frac{\mathscr{I}}{L})$ SUs should be implemented to sense the whole spectrum in one sensing period. Due to deep fading, J SUs are spatially implemented to sense the same L-channel group. Therefore, the jth SU implemented to sense the ith L-channel group is labeled as SU_{ij}. The whole scenario is shown in Fig. 4.8. In the second phase, a denoising algorithm is proposed to improve the detection performance of CSS, which is introduced in Sect. 4.4.2.

4.4.1 System Model

Based on the scenario shown in Fig. 4.8, the CSS algorithm based on low-rank MC can be formulated into a four-step model:

1. Sparse signals received at SUs.
2. Incomplete matrix generation at an FC.
3. MC at the FC.
4. Decision making at an FC.

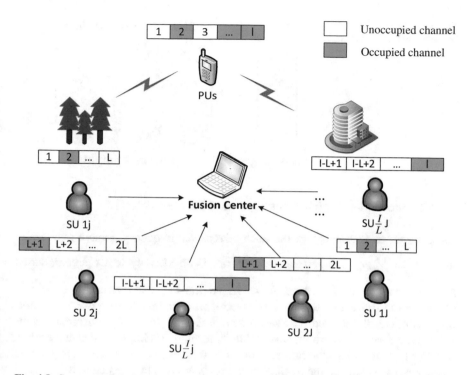

Fig. 4.8 System model of the proposed channel division scheme in cooperative spectrum sensing based on low-rank matrix completion

4.4.1.1 Signals Arrive at Secondary Users

As noise exists in the transmission channels, signals received at SU_{ij} are $r_{ij}(t) = h_{ij}(t) * s_{ij}(t) + w_{ij}(t)$, where $s_{ij}(t) \in \mathscr{C}^{n \times 1}$ refers to time domain signals of the ith L-channel group received at the jth receiver SU_{ij}, $h_{ij}(t)$ refers to the channel gain for the ith L-channel group between transmitter and SU_{ij}, $w_{ij}(t)$ refers to AWGN in the related transmission channels. The frequency domain representations of signals in the ith L-channel group which is received by SU_{ij} can be expressed as:

$$\mathbf{r_{fij}} = \mathbf{h_{fij}}\mathbf{s_{fij}} + \mathbf{w_{fij}}, \qquad (4.7)$$

where $\mathbf{r_{fij}}$, $\mathbf{h_{fij}}$, $\mathbf{s_{fij}}$, and $\mathbf{w_{fij}}$ are the DFT of $r_{ij}(t)$, $h_{ij}(t)$, $s_{ij}(t)$, and $w_{ij}(t)$.

At SU_{ij}, a random demodulator $\hat{\boldsymbol{\Phi}}_{ij} \in \mathscr{C}^{\tilde{p} \times n}$ is implemented to collect the compressed measurements as follows:

$$\mathbf{x_{ij}} = \hat{\boldsymbol{\Phi}}_{ij}\left(\mathbf{h_{fij}}\mathscr{F}^{-1}\mathbf{s_{fij}} + \mathscr{F}^{-1}\mathbf{w_{fij}}\right) \qquad (4.8)$$

$$= \hat{\boldsymbol{\Theta}}_{ij}\left(\mathbf{h_{fij}}\mathbf{s_{fij}} + \mathbf{w_{fij}}\right).$$

Fig. 4.9 Matrix to be recovered at the fusion center

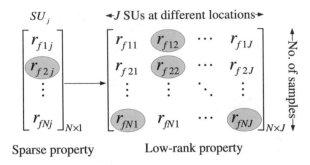

Sparse property   Low-rank property

4.4.1.2   Incomplete Matrix Construction at Fusion Center

As spectrum utilization is low, the stack of received frequency domain representations $\mathbf{r_{fj}} = \sum_{i=1}^{S} \mathbf{r_{fij}}$ are approximately sparse. Each SU only sends $\tilde{p}$ compressed measurements to an FC where $\tilde{p} < n$. At the FC, the matrix $\mathbf{R_f} \in \mathscr{C}^{N \times J}$ ($N = S \times n$) to be recovered shows a low-rank property transformed from the sparse property of signals received at SUs as shown in Fig. 4.9. In Fig. 4.9, the circled items refer to the observed measurements as the CS technique is implemented at each SU. In order to avoid poor spectrum resolution in frequency domain and high spectral leakage in each channel, the number of the rows N is set to be equal to the original number of samples for the whole spectrum of interest $S \times n$ rather than the number of channels $\mathscr{I}$, which is invoked in Wang et al. (2012); Ling and Tian (2011).

At the FC, only a subset $\Omega \subseteq \mathscr{C}^{M \times J}$ of $\mathbf{R_f}$ are collected where $P = S \times \tilde{p}$. Here, the compression ratio is defined as $\gamma = \frac{P}{N}$. We stack all columns $\mathbf{r_j}$ of $\mathbf{R_f}$ into a long vector as $\text{vec}(\mathbf{R_f})$. The incomplete matrix $\mathbf{X}$ is obtained by:

$$\text{vec}(\mathbf{X}) = \hat{\boldsymbol{\Theta}}\,\text{vec}(\mathbf{R_f}) = \hat{\boldsymbol{\Theta}}\,H\,\text{vec}(\mathbf{S_f}) + \hat{\boldsymbol{\Theta}}\,H\,\text{vec}(\mathbf{W_f})\,, \tag{4.9}$$

where $\mathbf{h_f} = \text{diag}(\text{diag}(h_{f11}, \dots, \mathbf{h_{f1J}}), \dots, \text{diag}(\mathbf{h_{f\mathscr{I}1}}, \dots, \mathbf{h_{f\mathscr{I}J}}))$, $H = \text{vec}(\mathbf{h_f})$, and $\hat{\boldsymbol{\Theta}} = \text{diag}\left\{\hat{\boldsymbol{\Theta}}_{11}, \dots, \hat{\boldsymbol{\Theta}}_{ij}, \dots \hat{\boldsymbol{\Theta}}_{\mathscr{I}J}\right\}$ is the block diagonal matrix. It is assumed $\mathbf{S_f}$ is the corresponding noiseless matrix of $\mathbf{R_f}$, and $\mathbf{W_f} = \mathbf{R_f} - \mathbf{S_f}$ is the matrix of the corresponding noise contained in $\mathbf{R_f}$. The unobserved measurements in $\mathbf{S_f}$ should be recovered from $\mathbf{X}$.

4.4.1.3   Matrix Completion at Fusion Center

The size of the matrix and the computational cost would increase when the number of the rows N is equal to the length of samples $S \times n$, which is the samples size of the whole spectrum of interest, to improve the frequency resolution. The signal recovery process is normally performed at SUs in the single node spectrum sensing algorithm, and SUs are normally power-limited devices Yucek and Arslan (2009). Therefore, the signal recovery process may cause long delay which will make the final decision invalid for the dynamic spectrum. However, in a CSS network, the MC process can be performed by a powerful device such as the FC to replace the power-limited SUs.

With the low-rank property, the complete matrix $\mathbf{S_f}$ can be recovered from a random subset of its elements in $\mathbf{X}$ at the FC. This MC problem is defined as Candes and Recht (2009)

$$\begin{aligned} &\min \ \mathrm{rank}\left(\hat{\mathbf{S}}_\mathbf{f}\right), \\ &\text{subject to } \left\|\hat{\boldsymbol{\Theta}} \cdot \mathbf{H}\mathrm{vec}(\hat{\mathbf{S}}_\mathbf{f}) - \mathrm{vec}(\mathbf{X})\right\|_2^2 \le \varepsilon, \end{aligned} \tag{4.10}$$

where $\hat{\mathbf{S}}_\mathbf{f}$ refers to the reconstructed matrix. Here, solving problem (4.10) refers to find a matrix with the minimal singular values but satisfies the constraints.

However, (4.10) is an NP-hard problem Candes and Recht (2009). It has been proved that such an NP-hard problem can be well approximated by nuclear norm minimization problem. Then, the MC problem can be formulated as

$$\begin{aligned} &\min \ \left\|\hat{\mathbf{S}}_\mathbf{f}\right\|_*, \\ &\text{subject to } \left\|\hat{\boldsymbol{\Theta}} \cdot \mathbf{H}\mathrm{vec}(\hat{\mathbf{S}}_\mathbf{f}) - \mathrm{vec}(\mathbf{X})\right\|_2^2 \le \varepsilon. \end{aligned} \tag{4.11}$$

Here, $\left\|\hat{\mathbf{S}}_\mathbf{f}\right\|_*$ refers to the sum of singular values of $\hat{\mathbf{S}}_\mathbf{f}$.

4.4.1.4   Decision Making at an Fusion Center

When the complete matrix $\hat{\mathbf{S}}_\mathbf{f}$ is obtained by solving (4.11), the average energy density of each L-channel group can be calculated and compared with the threshold λ defined in (2.11) to make the final decision on spectrum occupancy. Once the final decision is made, it should be sent back to each SU participating the cooperative networks to enable them getting access to the vacant channels.

4.4.2   Denoised Cooperative Spectrum Sensing Algorithm

Similarly to the denoising algorithm in the proposed robust single node spectrum sensing algorithm in (4.2), the bth sample in the recovered signal $\hat{s}_{fij}$ is compared with the corresponding noise level $\sigma_n(b)$. If $\hat{s}_{fij}(b)$ is higher than $\sigma_n(b)$, the measurement observed at the FC $r_{fij}(b)$ is kept. Otherwise, the corresponding value is set to be zero to eliminate the influence of noise. Here, the recovery error may be caused by the high channel noise or not enough number of compressed measurements. This process can be illustrated as:

$$\hat{s}_{fij\_d}(b) = \begin{cases} r_{fij}(b), & \text{if } \hat{s}_{fij}(b) \geq \sigma_n(b), \\ 0, & \text{otherwise.} \end{cases} \tag{4.12}$$

4.4.3 Computational Complexity and Performance Analyses

In the low-rank MC-based CSS scenario, the computational complexity of solving the MC problem is at the level of $O(N^3)$, and the MC is performed at a very powerful FC. As a result, the complexity introduced by MC would not be a key issue to be considered. In such a case, the key issue is the high sampling requirement for wideband spectrum at SUs with limited sensing capability.

In the proposed robust CSS algorithm based on low-rank MC, the bandwidth to be sensed at each SU is reduced to L out of $\mathscr{I}$ channels. Additionally, each SU performs sub-Nyquist sampling and only the collected samples p are sent to the FC which would lower the transmission load in the networks in comparison with the scenario where all the n samples are sent to an FC. Meanwhile, $\frac{\mathscr{I}}{L}$ SUs are needed to employ at different locations to sense the whole spectrum of interest. As the spatial diversity of SUs is utilized to avoid the deep fading problem in CSS network, the more SUs participating in the CSS network, the better detection performance can be achieved. In such a case, if each SU only senses part of the spectrum, detection performance will be degraded. This tradeoff is illustrated in the following simulations. In large-scale CRNs, such kind of performance degradation can be compensated as the number of participating SUs is large.

4.5 Numerical Analyses for Cooperative Multiple Users Case

4.5.1 Analyses on Simulated Signals

In the multiple node scenario, the spectrum of interest is TVWS with $\mathscr{I} = 40$ channels. Each SU is assumed to sense a non-overlapping L-channel ($L = 8$) group which is the same as the simulation setup of the single node spectrum sensing scenario in Sect. 4.3. The target P_f is set to be 0.01. Transmission channels between the transmitters to the SUs experience frequency-selective fading. In each sensing period, the fading on each channel is time-invariant and it is modeled by setting a random delay and independent Rayleigh fading gains for the multipath fading channels. Without loss of generality, the first SU participating in the cooperative networks is assumed to experience deep fading and the rest of SUs are experiencing Rayleigh fading. In the following simulations, the performance of proposed robust

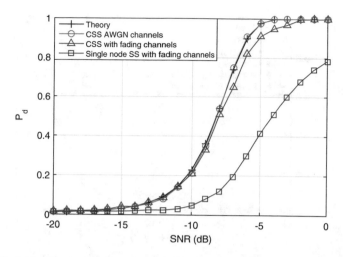

Fig. 4.10 P_d comparison of theoretic curves, CSS under AWGN channels and deep fading channels, and single node spectrum sensing under deep fading channels with simulated signals

CSS algorithm is presented by considering the influence of multipath deep fading, different number of measurements observed at the FC and different network sizes are analyzed.

The detection performance of single node spectrum sensing under deep fading channels in comparison with CSS algorithm with fading channels, AWGN channels, and the theoretical curves defined in (4.6) is shown in Fig. 4.10. It can be seen that P_d of the single node spectrum sensing, which can be considered as the number of SU implemented to sense each L-channel group is $J = 1$, becomes much lower than the theoretical curves when the transmission channels experience deep fading. As the spatial diversity gain of CSS, the detection performance of CSS algorithm is much improved even though the SUs experiencing deep fading are also in the cooperative network. In the CSS network, the number of SUs being implemented to sense each L-channel group is $J = 10$. It can be observed that the detection performance of CSS experiencing deep fading is still a bit lower than that of the theoretical curves and the CSS under AWGN channels.

Furthermore, it is noticed that the signal recovery step introduces most of the computational complexities among the four-step process for the single node spectrum sensing and the CSS algorithms. In the single node spectrum sensing algorithm based on CS, the signal recovery process is performed at the SU. However, in the CSS algorithm based on low-rank MC, signal recovery process is performed at the FC. SU devices, such as mobile phones and the slave WSDs, are normally battery powered Yucek and Arslan (2009) or even battery free for those nodes in WPT model in which the energy is harvested from power beacons. Therefore, the computation complexity should not be too high at the SUs. Otherwise, SUs cannot afford the sensing and signal recovery locally, and the delay caused by signal

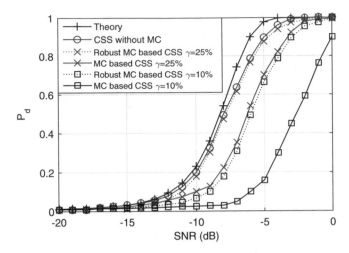

Fig. 4.11 P_d comparison of the proposed robust cooperative spectrum sensing algorithm with simulated signals under different compression ratios γ

recovery would be intolerable. As a result, the spectrum sensing decision may not be meaningful since spectrum occupancy may have changed during the period of signal recovery. However, for the FC, they are normally powerful devices such as base stations and master WSDs. In fact, size of the to be solved matrix at the FC is much greater than the number of samples to be recovered at SUs in the single node spectrum sensing, and the size of the matrix to be solved at the FC would also influence the performance of the proposed algorithm.

Figure 4.11 illustrates the detection performance comparison of the proposed robust CSS algorithm based on low-rank MC, low-rank MC-based CSS without denoising algorithm, CSS algorithm without CS technique implemented at SUs, and the theoretical values as defined in (4.6) under different number of observed measurements at the FC. In this scenario, the number of SUs being implemented to sense the same L-channel group is $J = 10$. The number of active PUs in each L-channel group is 1 with random position, corresponding to the sparsity level of 12.5% in the whole spectrum of interest, which is close to the real spectrum occupancy scenario Kolodzy and Avoidance (2002); Ofc (a). It is noticed the P_d increases when the number of observed measurements at the FC increases from 10 to 25%. As the MC error becomes lower with more observed measurements at the FC, the detection performance of proposed robust MC based CSS algorithm can almost match with that of CSS algorithm without CS implemented at SUs when the observed measurements at the FC is increased to 25%.

Figure 4.12 presents the P_d of the proposed robust CSS algorithm under different network sizes. In this scenario, the number of SUs being implemented to sense the same L-channel group is $J = 1, 2, 5, 10, 20$ and $J = 25$, respectively. In this scenario, the number of observed measurements at the FC is set to be 25% of the total measurements. With decreasing number of SUs participating in the CSS networks, the cooperative gain of CSS networks degrades. When the number of SUs

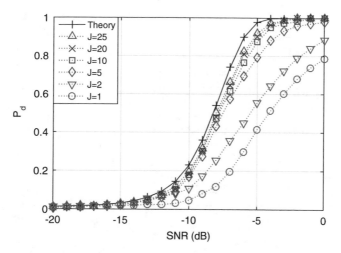

Fig. 4.12 P_d comparison of the proposed robust cooperative spectrum sensing algorithm with simulated signals under different network sizes, compression ratio $\gamma = 25\%$

implemented to sense the same L-channel group is decreased to $J = 1$, it becomes a single node spectrum sensing scenario, and the cooperative gain for CSS networks is decreased to zero. In such a case, it becomes a single node case which provides a benchmark for the comparison. It shows that the detection performance increases with increasing number of SUs implemented to sense the same L-channel group. It is also noticed that the performance gap for the number of SUs implemented to sense the same L-channel group increased from 5 to 10 is higher than that of the number of SUs changing from 10 to 20. As more information about the spectrum is sent to the FC for the final decision making, which refers to more SUs implemented to sense the same L-channel group, the detection performance becomes closer to the theoretical curves. However, when the network size is enlarged, the computational complexity of MC increases. Therefore, it is a balance between the detection performance and the computational complexity of MC. In addition, in the case $J = 5$, there are 25 SUs participating in the CSS network as each $\frac{\mathcal{J}}{L} = 5$ SUs are implemented to sense the whole spectrum of interest at the same location. It can be observed that the detection performance reaches the theoretic curves with increasing number of SUs. Therefore, the performance degradation caused by the proposed channel division scheme would not be an issue in large-scale networks.

4.5.2 Analyses on Real-World Signals

When the performance of proposed robust CSS algorithm based on low-rank MC is verified by the simulated signals, it is further tested on real-world signals collected by the RFeye sensing node installed in our lab as shown in Fig. 3.7 and a portable RFeye sensing node implemented at different locations in London.

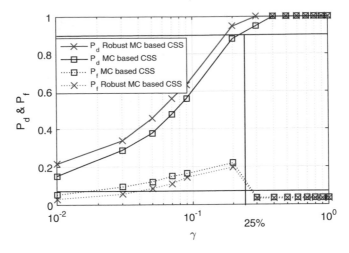

Fig. 4.13 Proposed robust cooperative spectrum sensing algorithm achieves higher P_d and lower P_f than the traditional algorithm with real-world signals under different compression ratios γ, and threshold λ is -73.5 dBm

Figure 4.13 shows the detection performance comparison of the traditional and the proposed robust CSS algorithms under different compression ratios when the real-world signals recorded by the RFeye node are utilized as the signal resources. In this scenario, the number of SUs used to sense the same channels is $J = 5$ and the threshold is set to be -73.5 dBm. It is noticed that detection performance of the proposed algorithm would reach the target performance (P_d is higher than 90% and P_f is lower than 10%) when the compression ratio γ is no lower than about 25%. In addition, the detection performance of the proposed robust CSS algorithm is better than the traditional one with increasing compression ratio at the SU, which is the benefit of the proposed denoising algorithm. When compared with Fig. 4.11, it can be observed that the detection performance becomes better when the compression ratio increases. This is also matched with the single node spectrum sensing algorithm based on CS in Fig. 4.7. Similar as the single node case shown in Fig. 4.7, there is a sharp dropping on P_f when the compression ratio γ becomes higher than 20%. The reason for the drop is that the MC becomes exact when the compression ratio is increased to be more than 20%, which lower the probability to falsely determine the unoccupied channels as occupied.

4.6   Summary

In this chapter, two algorithms for wideband spectrum sensing at sub-Nyquist sampling rates were proposed to reduce the computational complexity and improve the robustness to channel noise, which are designed for the cases of single SU and cooperative multiple SUs, respectively. The proposed algorithms were further tested on real-world signals after being validated by the theoretical results and the simulated signals. The analyses results showed that computational complexity of the proposed algorithms is much less than other state-of-the-art methods. Furthermore, simulation results demonstrated that the detection performance of the proposed spectrum sensing algorithms on both single and multiple nodes was more robust to channel noise than the traditional algorithms.

References

Akyildiz, I. F., Lo, B. F., & Balakrishnan, R. (2011). Cooperative spectrum sensing in cognitive radio networks: A survey. *Physical Communications, 4*, 40–62.

Bazerque, J., & Giannakis, G. (2010). Distributed spectrum sensing for cognitive radio networks by exploiting sparsity. *IEEE Transactions on Signal Processing, 58*, 1847–1862.

Candes, E., & Recht, B. (2009). Exact matrix completion via convex optimization. *Foundations of Computational Mathematics, 9*, 717–772.

Digham, F. F., Alouini, M.-S., & Simon, M. K. (2007). On the energy detection of unknown signals over fading channels. *IEEE Transactions on Communications, 55*, 21–24.

Ghasemi, A., & Sousa, E. (2005). Collaborative spectrum sensing for opportunistic access in fading environments. In *Proceedings of the IEEE International Symposium Dynamic Spectrum Access Networks (DYSPAN)*, Baltimore, MD (pp. 131–136).

Kolodzy, P., & Avoidance, I. (June 2002). Spectrum policy task force. In *Report ET Docket*. Washington, DC:Federal Communications Commission.

Li, Z., Yu, F. R., & Huang, M. (2010). A distributed consensus-based cooperative spectrum-sensing scheme in cognitive radios. *IEEE Transactions on Vehicular Technology, 59*, 383–393.

Ling, Q., & Tian, Z. (2011). Decentralized support detection of multiple measurement vectors with joint sparsity. In *Proceedings of IEEE International Conference on Acoustics, Speech, and Signal Processing (ICASSP)*, Prague, Czech (pp. 2996–2999).

Ma, Y., Gao, Y., Liang, Y. C., & Cui, S. (2016). Reliable and efficient sub-Nyquist wideband spectrum sensing in cooperative cognitive radio networks. *IEEE Journal on Selected Areas in Communications, 34*, 2750–2762.

Quan, Z., Cui, S., Sayed, A. H., & Poor, H. V. (2009). Optimal multiband joint detection for spectrum sensing in cognitive radio networks. *IEEE Transactions on Signal Processing, 57*, 1128–1140.

RFeye node https://www.crfs.com/all-products/hardware/nodes/.

Sun, H., Chiu, W.-Y., & Nallanathan, A. (2012). Adaptive compressive spectrum sensing for wideband cognitive radios. *IEEE Communications Letters, 16*, 1812–1815.

Tian, Z. (2008). Compressed wideband sensing in cooperative cognitive radio networks. In *Proceedings of the IEEE Global Communications Conference (GLOBECOM)*, New Orleans, LA (pp. 3756–3760).

Treichler, J., Davenport, M., & Baraniuk, R. (2009). Application of compressive sensing to the design of wideband signal acquisition receivers. *US/Australia Joint Workshop on Defense Applications of Signal Processing (DASP)* 5.

UK Office of Communications (Ofcom). (July 2009). *Statement on cognitive access to interleaved spectrum*.

Wang, Y., Tian, Z., & Feng, C. (2012). Collecting detection diversity and complexity gains in cooperative spectrum sensing. *IEEE Transactions on Wireless Communications, 11*, 2876–2883.

Wang, N., Gao, Y., & Zhang, X. (2013). Adaptive spectrum sensing algorithm under different primary user utilizations. *IEEE Communications Letters, 17*, 1838–1841.

Ye, Z., Memik, G., & Grosspietsch, J. (2008). Energy detection using estimated noise variance for spectrum sensing in cognitive radio networks. In *Proceedings of IEEE Wireless Communications and Networking Conference (WCNC)*, Las Vegas, NV (pp. 711–716).

Yucek, T., & Arslan, H. (2009). A survey of spectrum sensing algorithms for cognitive radio applications. *IEEE Communications Surveys & Tutorials, 11*, 116–130.

Zeng, F., Li, C., & Tian, Z. (2011). Distributed compressive spectrum sensing in cooperative multihop cognitive networks. *IEEE Journal of Selected Topics in Signal Processing, 5*, 37–48.

Zhang, X., Ma, Y., Qi, H., Gao, Y., Xie, Z., Xie, Z., et al. (2018). Distributed compressive sensing augmented wideband spectrum sharing for cognitive IOT. *IEEE Internet of Things Journal, 5*, 3234–3245.

Chapter 5
Secure Compressive Spectrum Sensing

In this chapter, a malicious user detection model is proposed to improve the security of CSS networks. The low-rank MC technique is invoked in the proposed model. More specifically, Sect. 5.1 introduces the related work and main contributions of the work in this chapter. Section 5.2 describes the system model of CSS networks with malicious users. Section 5.3 presents the proposed low-rank MC-based malicious user detection framework along with the proposed rank estimation algorithm and the estimation strategy for the number of malicious users. Section 5.4 shows the numerical analyses of the proposed framework on both simulated and real-world signals. Section 5.5 concludes this chapter.

5.1 Introduction

In CRNs, CSS is an effective approach to offer significant performance gain in incumbent detection by exploiting the spatial diversity of the collaborative SUs (Ghasemi and Sousa 2005; Akyildiz et al. 2011). However, due to the openness of low-layer protocol stacks, CSS networks are vulnerable to endure attacks from spectrum sensing data falsification (SSDF). This characteristic of CSS networks blocks the application of CR technique in large-scale networks.

In CSS networks, SUs that launch SSDF attackers are named as malicious users. The main goals of malicious attacks come from two aspects: (1) decreasing detection probability for disturbing the normal operations of PUs; (2) increasing false alarm probability to deprive access opportunities of the honest SUs (Ding et al. 2013). In decentralized CSS networks, sensing results are exchanged between neighbor SUs for improving the network reliability to link failure. However, this characteristic makes decentralized CSS more vulnerable to malicious attacks (Yan et al. 2012), as the observations at honest SUs would be known by malicious users during the convergence process. Furthermore, fake data can be integrated into the

© The Author(s), under exclusive license to Springer Nature Switzerland AG 2019
Y. Gao, Z. Qin, *Data-Driven Wireless Networks*, SpringerBriefs in Electrical
and Computer Engineering, https://doi.org/10.1007/978-3-030-00290-9_5

decisions of honest neighbor SUs, which eventually brings significant performance degradation of the whole CSS networks (Zhang et al. 2015). In centralized CSS networks, all SUs report their local sensing data to an FC, at which the final decision on spectrum occupancy is made. By doing so, all participating SUs including malicious users can only obtain the spectrum occupancy knowledge from the FC. Thus, the observations at honest SUs in CSS networks would not leak to malicious users directly. However, as the fake data are still considered in decision-making process, existence of malicious users may lead to false decisions at the FC. Generally, regardless of the types of malicious attacks and CSS networks, malicious users have posed significant challenges on the security in CSS networks. As a result, detection accuracy of malicious users is quite essential to guarantee the security of CSS networks.

Along with improving the security of CSS networks through malicious user detection, another key challenge for secure CSS networks comes from the data acquisition costs reduction at SUs. It is identified that this sparse property can be transformed to a low-rank property of the matrix constructed by spectral signals received at spatially distributed SUs (Qin et al. 2016; Ma et al. 2016, 2017), since nearby locations or adjacent channels are supposed to share the similar spectrum occupancies. The MC technique Candes and Recht (2009) can be applied to recover the complete matrix with only partial of observable elements. Specifically, by invoking MC technique at the FC, SUs in CSS networks can sense less number of channels as the unsensed channels can be reconstructed from the sensed channels based on the low-rank property.

5.1.1   Related Work

So far, malicious user detection has been widely researched for enhancing the security of CSS networks (Wang et al. 2009, 2010; Chen et al. 2008; Kaligineedi et al. 2008, 2010; Kalamkar et al. 2012; Li and Han 2010; Wang et al. 2014, 2015; Liu et al. 2017). Specifically, the performance of CSS networks with single and multiple malicious users was investigated in Wang et al. (2009, 2010), respectively. Particularly, based on the historical reports from SUs, the suspicious level of each SU as well as consistency values was calculated to alleviate the influences of malicious users. Chen et al. (2008) proposed a reputation-based mechanism to defense the malicious attacks. However, these historical data-based algorithms take a long time to build a reliable reputation. Additionally, Kaligineedi et al. (2010) proposed a robust outlier detection to identify "Always Yes" malicious users by utilizing outlier factors and spatial information of SUs. Kalamkar et al. (2012) proposed an outlier detection scheme to detect malicious users sending true or false power values randomly to confuse other SUs in CSS networks. Furthermore, some work has been done on the attacks detection from intelligent malicious users. Li and Han (2010) proposed an abnormality-detection approach for secure CSS networks, in which the attack strategy adopted by malicious users is unknown. Wang et al. (2014, 2015) constructed a moral hazard principal-agent framework for malicious user detection. More specifically, an incentive compatible mechanism was designed for thwarting the malicious behaviors from rational and irrational intelligent malicious users. By doing so, the proposed approach was more practical to be implemented in CR networks.

Besides the existing work on malicious user detection, the MC-based CSS networks have been studied in Meng et al. (2011), Wang et al. (2012a), Li (2010), with the purpose of alleviating the costs of data acquisition at SUs. Meng et al. (2011) firstly introduced the concept of MC to CSS networks. It was proposed that each SU linearly combined the information of multiple channels at sub-Nyquist sampling rates. Subsequently, each SU sent a small number of such linear combinations to an FC to perform MC. Additionally, Wang et al. (2012a) proposed a robust wideband spectrum sensing algorithm with sub-Nyquist sampling performed at each active SU in the considered CSS networks. Once the compressed measurements were sent to the FC, nuclear norm minimization was adopted to solve the low-rank MC problem. By doing so, the costs of data acquisition at SUs are reduced significantly. Furthermore, Li (2010) firstly applied belief propagation framework to MC for making it implementable and efficient on reconstructing spectrum occupancies in wideband CSS networks.

5.1.2 Motivations and Contributions

The aforementioned work has played a vital role and laid solid foundation for developing new strategies on malicious user detection. However, many of them are trust based, which utilizes the historical information of malicious users' behaviors. In practice, reliable reputation information is not always available since well-established historical statistics may be too expensive or even unrealistic in a fast-changing CR environment. Additionally, intelligent malicious users sending random values are more challenging than the types of malicious users considered in Kaligineedi et al. (2010), Kalamkar et al. (2012). Motivated by these, a malicious user detection dealing with malicious users sending random values is desirable for secure CSS networks. Another motivation of the work comes from reducing the number of active SUs in CSS networks and data acquisition costs at SUs without loss of any information. Here, the active SUs refer to SUs send data to the FC. All the aforementioned work on MC-based CSS focuses on reducing the costs of data acquisition at SUs without considering any secure issue in CSS networks. Therefore, a malicious user detection algorithm with energy efficiency at SUs is extremely challenging and desired.

In this chapter, compared to the CSS network without MC, a malicious user detection model with fewer number of active SUs is considered. To the best of my knowledge, this is the first work which invokes MC technique to achieve the malicious user detection in CSS networks. The contributions of this chapter are summarized as follows:

1. A malicious users detection framework is proposed without requiring prior information of networks. In the proposed framework, along with the reduction of the number of active SUs and the data acquisition costs at SUs, the accuracy of malicious user detection is improved.
2. In the proposed framework, compared to the CSS network without MC, fewer number of sensed channels is required, as MC technique is invoked at the FC to recover the information of unsensed channels. As a result, less number of active SUs are required. If there are enough active SUs in CSS networks, with the invoking of MC technique, each active SU can sense less number of channels without degrading the recovery performance. At the FC, sensed channels but corrupted by malicious users are removed during the MC process by utilizing the adaptive outlier pursuit (AOP) algorithm (Yan et al. 2013).
3. A dynamic rank estimation algorithm is proposed to provide the rank order as one of the inputs for AOP algorithm. By doing so, the proposed framework does not require any prior information of the considered CSS networks for malicious user detection at the FC.
4. An estimation strategy on the number of malicious users is proposed for the malicious user detection framework. With the new strategy, the estimated number of malicious users can be used as one of the inputs for AOP algorithm to make the malicious user detection algorithm completely blind.
5. The proposed framework is tested on the real-world signals after being validated by the simulated signals. Numerical results show that the proposed malicious user detection framework can achieve high detection accuracy with low costs of data acquisition at SUs or less number of active SUs.

5.2 System Model

5.2.1 Networks Description

We take a typical CSS scenario as the considered network model, as shown in Fig. 5.1a. It is assumed that the whole spectrum of interest with bandwidth $\mathscr{D}$ can be divided into $\mathscr{I}$ channels. A channel is either occupied by a PU or unoccupied. There is no overlap between different channels. The number of occupied channels K is assumed to be much less than the total number of channels, i.e., $K \ll \mathscr{I}$. Each channel is sensed by SUs at J different locations, which are spatially randomly distributed. At an arbitrary location indexed by j $(1 \leq j \leq J)$, it is assumed that B_j $\left(1 \leq B_j \leq \mathscr{I}\right)$ SUs, indexed by $\mathbf{b}$ $\left(1 \leq \mathbf{b} \leq B_j\right)$, are implemented to sense the spectrum of interest. In a conventional CSS network, the whole spectrum is sensed by an SU at each location, which results in $B_j = 1$. However, high sampling rates are challenging for SUs in a CSS network, as the SUs are normally energy-constrained with limited sensing capabilities.

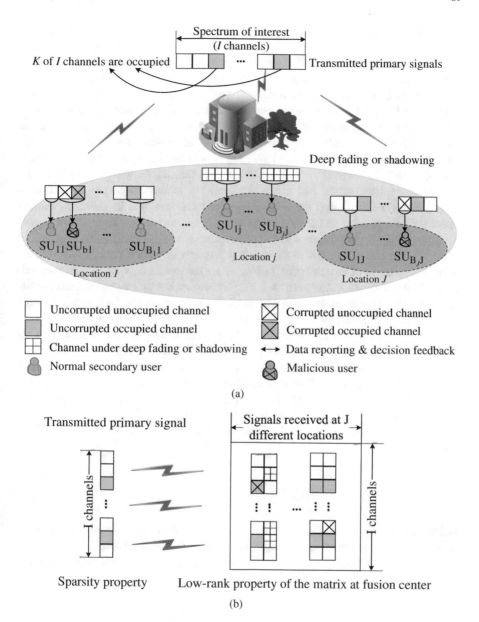

Fig. 5.1 Network model. (**a**) Network model of cooperative spectrum sensing network with malicious users. (**b**) Transformation of sparsity property into low-rank property

In this chapter, a few SUs are implemented at each location j of the CSS network ($B_j > 1$), where each SU only needs to sense a segment of the whole spectrum at Nyquist rates, which means that some of the channels are unsensed at one location as shown in Fig. 5.1a. Consequently, costs of data acquisition at SUs can be reduced significantly, in comparison with the case that each SU senses the whole spectrum. After sampling is performed, each SU calculates the power values of the sensed channels, and then sends this information to an FC to contribute to the final decisions on spectrum occupancies. It is further noticed that some of the SUs experience deep fading or shadowing. They would send very low power values to the FC in a CSS network, which are labeled as the blocks with "+" in Fig. 5.1a, b. The transmitted signal has a sparsity property in the frequency domain Tian and Giannakis (2007) and the nearby locations are assumed to share the similar spectrum occupancies Li (2010), so the matrix constructed by the received signals at different locations exhibits a low-rank property (Wang et al. 2012a). Figure 5.1b illustrates the transformation of the sparsity property of transmitted signals into the low-rank property of the matrix at the FC, where the matrix is constructed by signals received at different locations. In such a CSS network, we propose to reconstruct the unsensed channels from the sensed channels by a low-rank MC technique.

In the case of a sensing malfunction, some of the active SUs in the CSS network, labeled as the blocks with "X" in Fig. 5.1a, b, send corrupted power values to the FC. Malicious users appear randomly in the considered CSS network. Malicious users that keep sending high power values or low power values are easily detected. However, malicious users that send random but very close to the true values are much more difficult to detect. This is the case we consider in this work. We propose to remove these malicious users during the MC at the FC, so that recovery performance is not degraded significantly as the corrupted power values are used for the MC process.

5.2.2 Signal Processing Model

Let us define $s(t) \in \mathscr{C}^{N \times 1}$ as the transmitted signals from unknown PUs, where N refers to the number of samples. All active SUs in CSS networks are assumed to keep silent. Additionally, $r_{ij}(t)$ refers to the signals of the ith channel received at the jth SU (SU_{ij}), which can be given by

$$r_{ij}(t) = d_{ij}^{-\chi/2} h_{ij}(t) s(t), \tag{5.1}$$

where d_{ij} refers to the distance from PUs to SU_{ij} when the ith channel with response $h_{ij}(t)$ is used for transmission, and χ is the propagation loss factor.

Once signals of the ith channel are received at SU_{ij}, the power value of the sensed channel p_{ij} can be calculated as

$$p_{ij} = \frac{1}{N} \int_{f_i - \mathscr{D}/2\mathscr{I}}^{f_i + \mathscr{D}/2\mathscr{I}} \mathscr{F} r_{ij}(t) df, \tag{5.2}$$

where $\mathscr{F}$ denotes the DFT matrix. Defining $\boldsymbol{\Omega}$ as the index set for complete matrix at the FC, then the complete matrix $\mathbf{P}^{\Omega}$ can be illustrated as

$$
\mathbf{P}^{\Omega} = \begin{bmatrix} p_{1,1} & \cdots & p_{1,j} & \cdots & p_{1,J} \\ \vdots & \vdots & \vdots & \vdots & \vdots \\ p_{i,1} & \cdots & p_{i,j} & \cdots & p_{i,J} \\ \vdots & \vdots & \vdots & \vdots & \vdots \\ p_{I,1} & \cdots & p_{I,j} & \cdots & p_{I,J} \end{bmatrix}_{\mathscr{I} \times J} , \tag{5.3}
$$

where the ith row of $\mathbf{P}^{\Omega}$ represents the power values of the ith channel sensed by J SUs located spatially. The jth column of $\mathbf{P}^{\Omega}$ refers to the power values of different channels sensed by SUs at the jth location.

As each SU senses only one or a few channels among the whole spectrum of interest, power values collected at the FC $\mathbf{P}^{\mathbf{E}}$ are incomplete, where $\mathbf{E}$ is defined as an index set of the sensed channels at the FC. Therefore, power values of sensed channels in the incomplete matrix $\mathbf{P}^{\mathbf{E}}$ can be expressed as

$$
p_{ij}^{\mathbf{E}} = \begin{cases} p_{ij}, & (i, j) \in \mathbf{E}, \\ 0, & \text{otherwise.} \end{cases} \tag{5.4}
$$

If malicious users appear in CSS networks, part of the power values of the sensed channels would be corrupted during the data transmission from SUs to the FC. Let us define $\mathbf{O}$ as a subset of $\mathbf{E}$, which donates the sensed channels without corruption from malicious user. $\tilde{p}_{ij}$ is defined as the power value of corrupted channels. The value of $\tilde{p}_{ij}$ falls in the range of $[p_{min}^{\mathbf{EC}}, p_{max}^{\mathbf{EC}}]$, where $p_{min}^{\mathbf{EC}}$ and $p_{max}^{\mathbf{EC}}$ refer to the minimal and maximal values of the power values collected at the FC, respectively. Consequently, the partly corrupted matrix $\mathbf{P}^{\mathbf{EC}}$ generated at the FC can be expressed as

$$
p_{ij}^{\mathbf{EC}} = \begin{cases} p_{ij}, & (i, j) \in \mathbf{O}, \\ \tilde{p}_{ij}, & (i, j) \in \mathbf{E}/\mathbf{O}, \\ 0, & \text{otherwise,} \end{cases} \tag{5.5}
$$

5.3 Malicious User Detection Framework

In order to enhance the security of the CSS network, a malicious user detection framework based on low-rank MC is proposed in this section. Based on the network model described in Sect. 5.2.1, each SU is proposed to sense only a segment of the spectrum rather than the whole spectrum, in order to reduce the number of active SUs in the CSS network and the costs of data acquisition at each active SU. We propose to remove corrupted channels at the FC by invoking the AOP

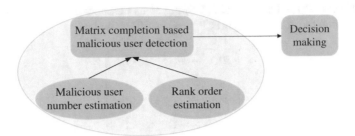

Fig. 5.2 Flowchart of the proposed malicious user detection framework with low-rank matrix completion

algorithm (Yan et al. 2013). It is further noted that the rank of the matrix at the FC and the number of channels corrupted by the malicious users are normally unknown in reality, but are required by the proposed malicious user detection algorithm with AOP. To make the proposed malicious user detection framework completely blind, a rank estimation algorithm and an estimation strategy on the number of malicious users are proposed in this section. As a result, the malicious user detection process in the CSS network does not require any prior information. Once the exact matrix is obtained by the proposed framework, spectrum occupancies can be determined by a conventional energy detection method. The whole procedure of the proposed low-rank MC-based malicious user detection framework in the CSS network is illustrated in Fig. 5.2.

5.3.1   Proposed Malicious User Detection Algorithm

As the spectrum is normally underutilized in reality, power values of all the channels received at SU_{ij} exploit a sparse property. This sparse property can be transformed into a low-rank property of the complete matrix $\mathbf{P}^{\Omega}$ at the FC. Therefore, the rank order of $\mathbf{P}^{\Omega}$ is equal to the number of active PUs in the spectrum of interest. Here, it is assumed that there is at least one active PU in the spectrum of interest, which guarantees the rank order is not equal to zero. Therefore, this low-rank property makes it possible to recover the unsensed channels at the FC by invoking MC technique.

As aforementioned, the sensed channels are partly corrupted by malicious users, which affects the recovery accuracy of sensed channels at the FC. It is assumed that the corrupted channels are distributed sparsely and randomly in the incomplete matrix at the FC. The indices of corrupted channels are unknown at the FC. Additionally, in order to make attacks more difficult to be detected, fake data corrupted by malicious users are assumed to be in a bounded range as aforementioned, which are close to their true values. These fake data can be removed

Fig. 5.3 Relationship illustration between the complete matrix $\mathbf{P}^{\Omega}$ and the corruption index matrix Λ

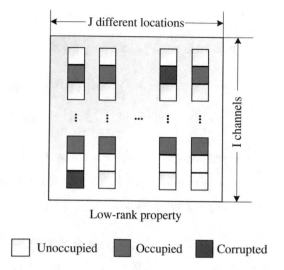

□ Unoccupied   ▨ Occupied   ■ Corrupted

during the process of MC by invoking the AOP algorithm. In such a case, the malicious user detection problem can be formulated as follows:

$$
\begin{aligned}
&\min_{\mathbf{U},\mathbf{V},\Lambda} \;\; \tfrac{1}{2} \sum_{(i,j)\in\Omega} \Lambda_{ij}\left((\mathbf{UV})_{ij} - p_{ij}^{\mathbf{EC}}\right)^2, \\
&\text{subject to} \;\; \sum_{(i,j)\in\Omega} \left(1 - \Lambda_{ij}\right) \le L_c, \quad \Lambda_{ij} \in \{0, 1\},
\end{aligned}
\tag{5.6}
$$

where $\mathbf{U} \in \mathscr{C}^{\mathscr{I} \times K}$ and $\mathbf{V} \in \mathscr{C}^{K \times J}$. The number of corrupted channels collected at the FC is L_c, and Λ is a binary matrix denoting the uncorrupted channels by one and the others are set to be zeros. An illustration for the structure of Λ is given in Fig. 5.3. More particularly, an arbitrary element Λ_{ij} in Λ is defined as

$$
\Lambda_{ij} = \begin{cases} 1, & (i, j) \in \mathbf{O}, \\ 0, & \text{otherwise.} \end{cases}
\tag{5.7}
$$

Here, $\mathbf{O}$ is updated in each iteration of solving problem (5.6). The details about how to update $\mathbf{O}$ will be given in Algorithm 1 in the following.

It is noted that the problem (5.6) is non-convex, since it has both continuous and discrete variables. The following two steps can be performed to find a local optimal solution to (5.6).

1. Fix Λ and update $\mathbf{U}$, $\mathbf{V}$. If $(i, j) \notin \mathbf{O}$, the objective function of (5.6) would become zero. Therefore, $\mathbf{U}$ and $\mathbf{V}$ can be obtained by solving the simplified problem as follows:

$$
\min_{\mathbf{U},\mathbf{V}} \sum_{(i,j)\in\mathbf{O}} \left((\mathbf{UV})_{ij} - p_{ij}^{\mathbf{EC}}\right)^2.
\tag{5.8}
$$

This problem can be easily solved by Riemannian trust-region for MC (RTRMC) Boumal and Absil (2011).

2. With fixed $\mathbf{U}$ and $\mathbf{V}$, Λ can be updated by solving

$$
\min_{\Lambda} \frac{1}{2} \sum_{(i,j)\in\Omega} \Lambda_{ij}\left((\mathbf{UV})_{ij} - p_{ij}^{\mathbf{EC}}\right)^2,
$$
$$
\text{subject to} \sum_{(i,j)\in\Omega} \left(1 - \Lambda_{ij}\right) \le L_c, \ \Lambda_{ij} \in \{0, 1\}. \tag{5.9}
$$

The problem (5.9) is to choose $(\mathscr{I} \times J - L_c)$ elements with least sum from $S_\Omega = \left\{\left((\mathbf{UV})_{ij} - p_{ij}^{\mathbf{EC}}\right)^2, (i, j) \in \Omega\right\}$. Given τ as the L_cth largest element in S_Ω, Λ_{ij} can be updated as

$$
\Lambda_{ij} = \begin{cases} 1, & (i, j) \in \Omega, \left((\mathbf{UV})_{ij} - p_{ij}^{\mathbf{EC}}\right)^2 < \tau, \\ 0, & \text{otherwise.} \end{cases} \tag{5.10}
$$

If the L_cth and $(L_c + 1)$th largest elements in S_Ω are equal, we can choose any Λ such that $\sum_{(i,j)\in\Omega} \left(1 - \Lambda_{ij}\right) = L_c$. Meanwhile, $S_{\Omega/\mathbf{O}} \ge S_\mathbf{O}$ should be satisfied, where $S_{\Omega/\mathbf{O}} = \min_{(i,j)\notin\mathbf{O}} \left((\mathbf{UV})_{ij} - p_{ij}^{\mathbf{EC}}\right)^2$ and $S_\mathbf{O} = \min_{(i,j)\in\mathbf{O}} \left((\mathbf{UV})_{ij} - p_{ij}^{\mathbf{EC}}\right)^2$, respectively.

During the process of solving problem (5.6), corrupted channels are removed from the observed ones at the FC to alleviate the influences of malicious users. Once the complete matrix $\hat{\boldsymbol{P}}^\Omega$ is recovered at the FC, final decision on spectrum occupancies can be determined by invoking the conventional energy detection. The ith channel is determined as occupied if the average energy of it is higher than the empirical threshold $\lambda_d = (\mu/2)^2$, where $\mu = \left\|\mathrm{vec}\left(\hat{\boldsymbol{P}}^\Omega\right)\right\|_1 / \left\|\mathrm{vec}\left(\hat{\boldsymbol{P}}^\Omega\right)\right\|_0$ is the average absolute value of all the $J \times K$ non-zero elements in $\mathrm{vec}\left(\hat{\boldsymbol{P}}^\Omega\right)$. Here, $\mathrm{vec}(\cdot)$ stacks all columns of matrix $\hat{\boldsymbol{P}}^\Omega$ into a long vector. The final binary decisions $\boldsymbol{d} = \{d_i, \forall i = 1, \ldots, \mathscr{I}\}$ on spectrum occupancies can be determined as

$$
d_i = \left(\frac{1}{J} \sum_{j=1}^{J} |\hat{p}_{ij}| \ge \lambda_d\right), \ \forall i, \tag{5.11}
$$

where $\hat{p}_{ij}$ is recovered power value of the ith channels sensed by SU_{ij}.

The proposed AOP-based malicious user detection is summarized in Algorithm 1. In the considered model, if the whole spectrum of interest are all sensed by SUs at different locations, the number of sensed channels at the FC is $P = \mathscr{I} \times J$. In such as case, a complete matrix is available at the FC. Otherwise, only an incomplete matrix can be constructed at the FC with $P < \mathscr{I} \times J$. Herein, the compression ratio

---

Algorithm 1 Proposed Malicious User Detection by Adaptive Outlier Pursuit

---

Require: Ω, P, $I_{\max}$, L, K, and λ_d.
Ensure: $l = 0$, $\Lambda_{ij} = 1$ for $(i, j) \in \Omega$, $\mathbf{O} = \Omega$.
 1: **while** $l \leq I_{max}$ **do**
 2:     Update $\mathbf{U}^{(l)}$ and $\mathbf{V}^{(l)}$ with RTRMC as in (5.8);
 3:     Update $\boldsymbol{\Lambda}^{(l)}$ with (5.10);
 4:     Update $\mathbf{O}^{(l)}$ to be indices in Ω where $\Lambda_{ij}^l = 1$;
 5:     **if** $\mathbf{O}^{(l)} = \mathbf{O}^{(l-1)}$ **then**
 6:         break;
 7:     **end if**
 8:     $l = l + 1$.
 9: **end while**
10: $\hat{p}_{ij} = \left(\mathbf{U}^{(l)} \mathbf{V}^{(l)} \right)_{ij}$.
11: Making final decisions $\boldsymbol{d}$ on spectrum occupancy by (5.11).
12: **return** $\boldsymbol{d}$.

$\gamma = \frac{P}{\mathscr{I} \times J}$ is defined as the ratio of number of the sensed channels P in the corrupted matrix $\mathbf{P}^{\mathbf{EC}}$ to the total number of channels $\mathscr{I} \times J$ in the complete matrix $\mathbf{P}^{\Omega}$ at the FC. Additionally, the maximal number of iterations for solving (5.6) is predefined as $I_{\max}$, which means the iterative process for solving (5.6) will be terminated when iteration number reaches $I_{\max}$, even though the break condition of Algorithm 1 is not satisfied. In the lth iteration, $\boldsymbol{\Lambda}^{(l)}$ is used to identify the locations of corrupted channels based on the newly constructed $\left(\hat{\boldsymbol{P}}^{\Omega} \right)^{(l)} = \mathbf{U}^{(l)} \mathbf{V}^{(l)}$. After the complete matrix is recovered, final decisions $\boldsymbol{d}$ on spectrum occupancies are made according to (5.11).

As demonstrated in Algorithm 1, the rank order of the matrix K and the number of corrupted channels L_c are required as the inputs at the FC. Therefore, a rank order estimation algorithm and an estimation strategy on the number of malicious users are proposed, as introduced in the following, to enable the proposed malicious user detection framework.

5.3.2   Rank Order Estimation Algorithm

It is pointed out that the rank order estimation can be converted into a sparsity order estimation problem (Wang et al. 2012a). With the sensed channels at the FC, the rank order of a matrix can be estimated by the following two steps:

1. Recover the complete matrix by solving

$$
\min \left\| \mathrm{vec} \left(\hat{\boldsymbol{P}}^{\Omega} \right) \right\|_1,
$$
$$
\text{subject to } \mathscr{A} \left(\hat{\boldsymbol{P}}^{\Omega} \right) = \mathbf{P}^{\mathbf{EC}},
\tag{5.12}
$$

where $\mathscr{A}$ is an operator from $\boldsymbol{\Omega}$ to $\boldsymbol{\Omega}/\mathbf{E}$. Here, the sparsity level of vec $\left(\hat{\boldsymbol{P}}^{\boldsymbol{\Omega}}\right)$ is equal to $J \times K$.

2. The estimated rank order $\hat{K}$ is given by

$$
\hat{K} = \sum_{i=1}^{\mathscr{I}} \left(\left| \frac{1}{J} \sum_{j=1}^{J} \hat{p}_{ij} \right| \geq \lambda_r \right), \tag{5.13}
$$

where λ_r is a threshold for rank order estimation. By applying data fusion at the FC, the power value of each channel is calculated by averaging power values of the same channel sensed by spatially distributed SUs. During the rank order estimation process, it is assumed that the existence of malicious users in CSS networks would not influence the rank order, as they are distributed randomly. This assumption can be guaranteed by the AOP algorithm Yan et al. (2013).

It is proved that exact signal recovery can be guaranteed when the number of sensed channels satisfies $P = c\,(K \times J) \log\left(\mathscr{I}/K\right) + d$ (Candes 2006). However, the number of sensed channels guaranteeing exact rank order estimation and exact MC is different. When Monte Carlo simulations and curve fitting techniques are adopted to find the values of the constants c and d, the following two results can be obtained with given $\mathscr{I}$, J, and K:

Result 1 Wang et al. (2012b) A successful rank order estimation can be guaranteed when the number of sensed channels is not less than P_1 where

$$
P_1 = c^1 (K_{\max} \times J) \log\left(\mathscr{I}/K_{\max} + d^1\right), \tag{5.14}
$$

where K_{max} is the statistical upper bound of the rank order K.

Result 2 Wang et al. (2012b) A successful MC can be guaranteed when the number of sensed channels is not less than P, which is defined as follows:

$$
P = c^2 (K \times J) \log\left(\mathscr{I}/K + d^2\right). \tag{5.15}
$$

According to the results presented in Wang et al. (2012b), it always holds that $P > P_1$ with given $\mathscr{I}$, J, and K. Therefore, $c^1 < c^2$ and $d^1 < d^2$.

Normally, the maximal rank order $K_{\max}$ is adapted as a statistical upper bound of the real rank order K. In practice, spectrum occupancies are normally dynamic. Therefore, $K_{\max}$ would not be a suitable upper bound in a dynamic spectrum environment. In order to obtain the exact rank order of the matrix at the FC, a novel dynamic rank order upper bound adjustment scheme is proposed to adjust $K_{\max}$ adaptively. One possible scenario is that $K_{\max}$ is much larger than K, which leads to that the number of data collected for rank order estimation is more than that for MC. Here, it is a waste on the costs of data acquisition at SUs or the number of active SUs implemented for sensing. As shown in Algorithm 2, the rank order upper

---

Algorithm 2 Proposed Shrink Algorithm

Ensure: $\mathbf{P}^{EC}$, Δ_1, $K_{\max}$, λ_r, and $J_{\max}$.
Require: $j = 1$, $P_2^{(1)} < 0$, $K_{\max}^{(1)} = K_{\max}$.
1: **while** $P_2^{(j)} < 0$ or $j \leq J_{\max}$ **do**
2:    Calculate $P_1^{(j)}$ by (5.14);
3:    Calculate $\hat{K}^{(j)}$ by (5.12) and (5.13) with $P_1^{(j)}$ channels;
4:    Calculate $P^{(j)}$ with $\hat{K}^{(j)}$ by (5.15);
5:    Calculate $P_2^{(j)} = P^{(j)} - P_1^{(j)}$;
6:    Update $K_{\max}^{(j)} = K_{\max}^{(j)} - \Delta_1$ and $j = j + 1$.
7: **end while**
8: **return** Updated $K_{\max}$.

---

bound adjustment can be achieved by the proposed shrink algorithm. $J_{\max}$ is defined the maximal iteration number which is adopted in Algorithm 2 to update $K_{\max}$. It is assumed that the sensed channels $\mathbf{P}^{EC}$, step size Δ_1, threshold λ_r, $J_{\max}$, and the initial rank order upper bound $K_{\max}$ are known as the inputs at the FC. Then the value of P_1 by (5.14) can be calculated. Additionally, the complete matrix can be obtained with the P_1 sensed channels by (5.12), and $\hat{K}$ can be determined by (5.13). Furthermore, the number of sensed channels required for exact recovery is obtained by (5.15). Subsequently, P_2 is updated as $P_2 = P - P_1$. In the next loop, $K_{\max}$ is reduced by step size Δ_1 and this process is repeated until $P_2 > 0$ or the maximal iteration $J_{\max}$ is researched.

Another scenario is that the rank order upper bound $K_{\max}$ is much smaller than the real rank order K, which is caused by the over-utilizing of shrink algorithm or the dynamic spectrum occupancies. It leads to the result that P_1 is not enough for the exact rank order estimation $\hat{K}$. As shown in Algorithm 3, the enlargement algorithm is proposed to enlarge $K_{\max}$ until that the exact rank order estimation can be achieved. In the proposed enlargement algorithm, sensed channels $\mathbf{P}^{EC}$, step size Δ_1, threshold λ_r, tolerance ε, and the initial rank order upper bound $K_{\max}$ are known as inputs at the FC. Subsequently, P_1 is determined by (5.14) to achieve exact rank order estimation. Additionally, $\hat{K}$ is determined by (5.12) and (5.13). In the following loop, $K_{\max}$ is increased by step size Δ_2 to get the updated $\hat{K}$ until the difference between $K_{\max}$ and $\hat{K}$ becomes less the error tolerance ε or the maximal iteration $J_{\max}$ is researched.

The whole process of the proposed dynamic rank order upper bound adjustment scheme is summarized as follows. As shown in Fig. 5.4, sensed channels $\mathbf{P}^{EC}$, step size Δ_1, threshold λ_r, tolerance ε, $J_{\max}$, and the initial rank order upper bound $K_{\max}$ are the inputs of the proposed scheme. The updated $K_{\max}$ is as the output. Once the inputs are available, the proposed shrink algorithm starts working. The output of Algorithm 2 $K_{\max}$ is adopted as one of the inputs for Algorithm 3. During this process, exact recovery cannot be achieved if the number of sensed channels P is smaller than P_1. As such, the FC should coordinate the number of active SUs in CSS networks to sense more channels. After the enlargement algorithm is

---

Algorithm 3 Proposed Enlargement Algorithm

Ensure: $\mathbf{P}^{EC}$, Δ_2, K_{max}, λ_r, ε and J_{max}.
Require: $j = 1$, $\hat{K}^{(1)} = 0$, $\hat{K}_{max}^{(1)} = K_{max}$.
1: **while** $\left(K_{max}^{(j)} - \hat{K}^{(j)} \right) > \varepsilon$ or $j \le J_{max}$ **do**
2:     Update $K_{max}^{(j)} = K_{max}^{(j)} + \Delta_2$ and $j = j + 1$;
3:     Calculate $P_1^{(j)}$ by (5.14);
4:     Calculate $\hat{K}^{(j)}$ by (5.12) and (5.13) with $P_1^{(j)}$ channels.
5: **end while**
6: **return** Updated K_{max}.

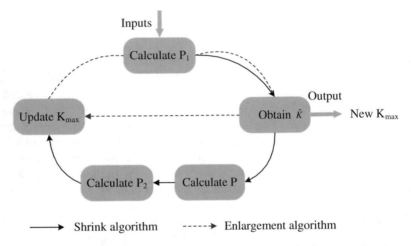

Fig. 5.4 Flowchart of the proposed dynamic rank order upper bound adjustment scheme

performed, the updated K_{max} can be obtained and used as the rank order input K for Algorithm 1. According to the logic flow shown in Fig. 5.4, the step size should be carefully designed. If the step size is too small, more iterations are required before the algorithm converges, which results in high computational complexity. If the step size is too big, K_{max} might keep updating by Algorithms 2 and 3 until the maximal iteration J_{max} is reached.

5.3.3   Malicious User Number Estimation

As aforementioned, the number of corrupted channels L_c is one of the inputs for Algorithm 1. However, it is usually unknown and needs to be estimated in practice. In the rest of this section, each SU is assumed to sense one channel to simplify the description. The number of malicious user is equal to the number of corrupted channels. Therefore, the malicious user number estimation refers to estimate the number of corrupted channels in the following of this chapter.

If the estimated malicious user number $\hat{L}_c$ is smaller or greater than its real value L_c, the performance of Algorithm 1 would be degraded significantly. More specifically, if $\hat{L}_c$ is underestimated, part of the corrupted channels will still be used to perform MC at the FC, which results in recovery errors in the reconstructed matrix. If $\hat{L}_c$ is overestimated, some of the uncorrupted channels would be removed during the MC process. The MC process would result in more than one solution. Consequently, exact MC is difficult to achieve as the number of available uncorrupted channels may not be enough. As proved in Yan et al. (2013), a sufficient condition for the non-uniqueness of a matrix $\mathbf{P}^{\Omega}$ is given as follows: suppose the number of sensed channels is P, and they are randomly distributed among the complete matrix $\mathbf{P}^{\Omega} \in \mathscr{C}^{\mathscr{I} \times J}$. Let us define ΔL_c as the difference between the overestimated number of corrupted channels $\hat{L}_c$ and the real number of corrupted channels L_c. If $\Delta L_c > (P - L_c)/\max(\mathscr{I}, J) - K > 0$, then the reconstructed matrix is non-unique.

In the proposed framework, after the rank order estimation is performed, an initial guess for malicious user number is used as one input for the AOP algorithm illustrated as Algorithm 1. It is an iterative process to update the number of malicious users in combining with the AOP algorithm. In each iteration, after the AOP algorithm is performed, the value of the $\hat{L}_c$th largest term in set S_{Ω} should be checked. If it is less than the tolerance l_{tol}, $\hat{L}_c$ is determined as overestimated, and some of the removed channels are uncorrupted. The numerical analyses in Yan et al. (2013) have proven that ΔL_c can be bounded by $(l_{\min} - K)$, where $l_{\min}$ is defined as the minimum number of sensed channels in one row or one column of the incomplete matrix with $(P - L_c)$ elements at the FC. More specifically, let us define $\hat{l}_{\min}$ as the minimal number of sensed channels in one row or one column of the incomplete matrix with $\left(P - \hat{L}_c\right)$ elements. If $\hat{l}_{\min}$ is less than the rank order K, the estimated number of corrupted channels $\hat{L}_c$ is updated as $\hat{L}_c = \hat{L}_c + \hat{l}_{\min} - K$. If $\hat{l}_{\min}$ is no less than K and $\left(\left(\mathbf{P}^{\Omega}\right)_{ij} - p_{ij}^{\text{EC}}\right)^2$ is less than τ, the exact matrix is reconstructed. Consequently, the iterative process for MC is terminated. Otherwise, if $\hat{l}_{\min}$ is greater than K and $\left(\left(\mathbf{P}^{\Omega}\right)_{ij} - p_{ij}^{\text{EC}}\right)^2$ is greater than τ, $\hat{L}_c$ is considered to be underestimated. As a result, the value of $\hat{L}_c$ should be updated to be $\rho_1 \hat{L}_c$, where $\rho_1 > 1$ is a properly selected constant. Following this, the updated $\hat{L}_c$ is taken as the input for AOP algorithm in the next iteration until the exact matrix is obtained or the iteration number reaches its upper bound $I_{\max}$.

5.3.4 Analyses on Minimal Number of Active Secondary Users

As aforementioned, each SU can sense one or multiple channels depending on its sensing capability. To simplify the comparison, it is assumed that each SU only senses one of the $\mathscr{I}$ channels. Without the invoking of MC technique, regardless of malicious users, the total number of SUs to be implemented in the CSS networks is $C_1 = \mathscr{I} \times J$. Additionally, with the invoking of MC at the FC, a CSS network without malicious users is considered. Without loss of any cooperative gain, the minimal number of SUs to be implemented in the CSS networks is given by

$$C_2 = \gamma_{\min} \times \mathscr{I} \times J, \qquad (5.16)$$

where $\gamma_{\min} \in (0, 1]$ is the lower bound of compression ratio for exact MC, which is dependent on the specific MC algorithm. Furthermore, based on the CSS networks considered in this chapter, in which malicious users exist and the AOP algorithm is invoked for MC, the minimal number of SUs required to be implemented in the CSS networks can be given by

$$C_3 = P = \hat{\gamma}_{\min} \times \mathscr{I} \times J, \qquad (5.17)$$

where $\hat{\gamma}_{\min} \in [\gamma_{\min}, 1]$ is the minimal compression ratio that can be achieved by the AOP algorithm. The exact value of $\hat{\gamma}_{\min}$ is dependent on the malicious user ratio κ. When the malicious user ratio $\kappa = 0$, $C_3 = C_2 < C_1$. If $\kappa > 0$, $C_2 < C_3 < C_1$. No matter whether the malicious users exist in CSS networks, with MC invoked at the FC, the number of active SUs required to send data to the FC is less the case that MC is not invoked.

Besides reducing the number of active SUs in the considered CSS networks, the costs of data acquisition can be reduced significantly from another perspective. Here, it is assumed that each SU senses multiple channels. With the proposed malicious user detection framework, if the number of active SUs is fixed, each SU can sense less number of channels as the MC technique is invoked to recover the complete matrix with less number of sensed channels. By sensing less number of channels, the costs of data acquisition at each SUs can be lowered significantly.

5.4 Numerical Analyses

In this section, numerical analyses on the proposed malicious user detection framework are presented. Particularly, the proposed framework is tested on the real-world signals after being verified by the simulated signals over TVWS.

In simulations, the total number of channels in the spectrum of interest is assumed to be $\mathscr{I} = 40$, which is the number of TVWS channels in the UK. The size of CSS networks changes from small scale ($J = 40$) to large scale ($J = 400$). Here, the size of CSS networks is equal to the number of active SUs implemented to sense the same channel at different spatial locations. Additionally, the malicious user ratio $\kappa = \frac{L_c}{\mathscr{I} \times J}$ is defined as the ratio of the number of corrupted channels L_c to the total number of channels ($\mathscr{I} \times J$) to be sensed by different SUs in the considered CSS networks. $I_{\max} = 500$ and $I_{\max} = 10$.

5.4.1 Numerical Results Using Simulated Signals

5.4.1.1 Results of the Proposed Rank Order Estimation

The simulation results of the proposed rank order estimation algorithm are presented in Fig. 5.5, with varying spectrum occupancies. In this scenario, dynamic spectrum occupancies result in changing rank order of the matrix at the FC. As shown in Fig. 5.5, when the rank order K changes, it can be observed that the saved number of channels to be sensed for exact MC is degraded at the changing point. However, performance of the proposed rank order estimation algorithm would be improved after the sensing slot during which rank order K changes. In practice, spectrum occupancies are assumed to be the same within a limited period. Therefore, the proposed rank order estimation algorithm is reliable for practical scenario. As shown in Fig. 5.5, with longer sensing period, the proposed rank order estimation algorithm outperforms the traditional two-step compressive spectrum sensing algorithm (TS-CSS) Wang et al. (2012b) in terms of the saved number of channels to be sensed for exact MC.

5.4.1.2 Results of the Case with Unknown Number of Malicious Users

In Fig. 5.6, the influences of incorrect estimation on the number of malicious users $\hat{L}_c$ are analyzed. Therefore, the proposed estimation strategy on the number of malicious users is not invoked here. In this case, we choose $J = 40$ and $K = 1$ to simplify the simulation process. Additionally, the estimated number of corrupted

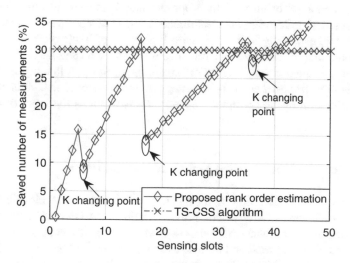

Fig. 5.5 Saved measurements for exact MC with dynamic spectrum occupancies at the fusion center

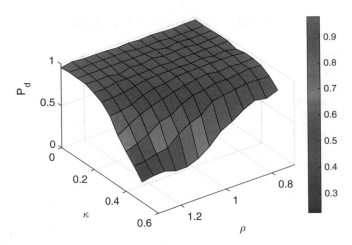

Fig. 5.6 Detection performance of the proposed malicious user detection framework versus different estimation accuracy ratio ρ and malicious user ratios κ, compression ratio $\gamma = 100\%$, and $J = 40$

channels $\hat{L}_c = \rho L_c$ varies from $0.7L_c$ to $1.3L_c$. Here, ρ, named as estimation accuracy ratio, is defined as the ratio of estimated number of corrupted channels to the actual number of corrupted channels. Here, the compression ratio is set to be 100%, which refers to the case that no MC is adopted. This kind of setting is to eliminate any possible performance degradation caused by recovery error. As a result, the performance difference shown in Fig. 5.6 is only caused by the incorrect estimation on the number of malicious users. Particularly, in Fig. 5.6, it is shown that the detection probability of the proposed malicious user detection framework gets degraded significantly if the estimated number of corrupted channels $\hat{L}_c$ is overestimated, especially in the case with high level of malicious user ratio. It is further noted that the detection performance would only be degraded slightly if the number of corrupted channels $\hat{L}_c$ is underestimated. In the following simulations, by invoking the proposed estimation strategy on the number of malicious users, the correct estimation of the corrupted channels $\hat{L}_c$ is taken as the input of the AOP algorithm.

5.4.1.3 Results of the Proposed Malicious User Detection

Figure 5.7 shows the detection performance of the proposed malicious user detection framework versus different compression ratios γ. Additionally, the detection performance of a traditional CSS network is also shown as a benchmark, in which malicious users do not exist and no MC technique is invoked at the FC. Therefore, for the case of the traditional CSS networks, matrix observed at the FC is complete. In this scenario, the number of active PUs in the spectrum of interest is one, which results in the rank order as $K = 1$. The number of active SUs implemented to sense

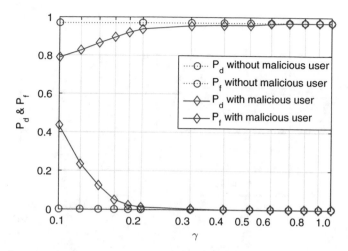

Fig. 5.7 Detection performance of the proposed malicious user detection framework versus different compression ratios γ, malicious user ratio $\kappa = 10\%$, and $J = 40$

each channel is assumed to be $J = 40$. As shown in Fig. 5.7, with 10% of sensed channels corrupted by malicious user, the detection probability and false alarm probability of the proposed framework can almost match with the benchmark, when the compression ratio γ is increased to 30%. This means the detection performance of the CSS networks would not be degraded if the number of sensed channels is no less than 30% of the total number of channels $\mathscr{I} \times J$, even though 10% of them are corrupted by malicious users.

Figure 5.8 shows the detection performance of the proposed malicious user detection framework versus varying malicious user ratios κ. In this case, the number of active PUs is set to be $K = 1$. The total number of active SUs implemented to sense each channel is $J = 40$. It is noted that detection probability of the proposed malicious user detection algorithm decreases with increasing number of channels corrupted by malicious users. More specifically, when malicious user ratio is increased to 60%, detection probability of the proposed framework is heavily degraded regardless of the compression ratio. It is reasonable because that the number of uncorrupted channels is not enough to guarantee the exact MC at the FC.

Figure 5.9 shows the detection performance of the proposed malicious user detection framework versus different number of active SUs J for sensing the same channel. Here, different number of active SUs for sensing the same channel leads to different sizes of CSS networks. In this scenario, the active of active PUs is set to be $K = 4$ with random positions. The size of CSS networks J varies from 40 to 400. The malicious user ratio varies from 10% to 60%. If the malicious user ratio is fixed, the number of corrupted channels would increase accordingly with the increasing size of CSS networks. Therefore, as illustrated in Fig. 5.9, with the same malicious user ratio, the case with larger CSS network size may have worse

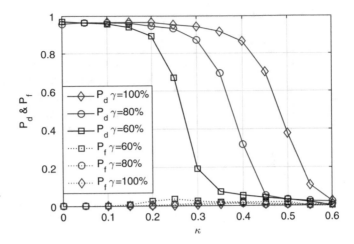

Fig. 5.8 Detection performance of the proposed malicious user detection framework versus different compression ratios γ, malicious user ratio κ varies from 0% to 60%, and $J = 40$

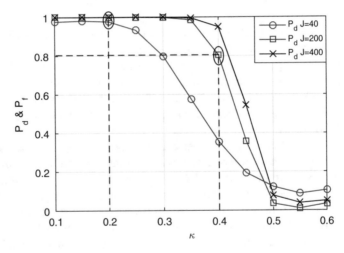

Fig. 5.9 Detection performance of the proposed malicious user detection framework versus different sizes of CSS networks J, malicious user ratio κ changes from 10% to 60%, and compression ratio $\gamma = 100\%$

detection performance than the case with smaller CSS networks. Additionally, as circled in Fig. 5.9, for the scenarios of $J = 200$ and $J = 400$, the number of corrupted channels becomes the same if the malicious user ratio is set to be 0.4 and 0.2, respectively. In such a case, it can be also noticed that detection performance of CSS networks with $J = 400$ is much higher than that with $J = 200$. Therefore, it can be concluded that the more number of active SUs in CSS networks (larger CSS network size), the stronger defense to the same number of malicious users.

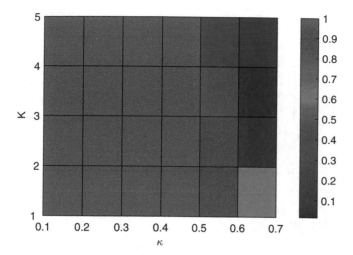

Fig. 5.10 Detection performance of the proposed malicious user detection framework versus different rank orders K and different malicious user ratios κ, compression ratio $\gamma = 100\%$, $J = 400$

In Fig. 5.10, detection performance of the proposed malicious user detection framework is presented versus different malicious user ratios κ and different rank orders K. As aforementioned, the rank order of the matrix at the FC is determined by the number of active PUs in the spectrum of interest. Positions of the active PUs are randomly generated in the spectrum of interest. In this case, compression ratio is set to be 100% to avoid any possible performance degradation caused by not enough number of sensed channels at the FC. The number of active SUs to sense each channel is set to be $J = 400$. It shows that the detection performance can be improved accordingly with decreasing rank order as well as decreasing malicious user ratio. This observation is reasonable, as increasing rank order and malicious user ratio would make the exact MC more difficult or even impossible at the FC.

5.4.2 Numerical Results Using Real-World Signals

As aforementioned, Ofcom has conducted serial trials on the TVWS pilots. One of the trials is conducted in our campus. In this trial, the DVB-T signal is allowed to be transmitted over TVWS channel 27 (518–526 MHz), which is used to be vacant. During this trial, the real-world signals over TVWS are collected by a portable CRFS RFeye node RFe as shown in Fig. 3.7. To simulate the CSS networks with malicious users, the signal transmitted in channel 27 is regarded as the malicious users over TVWS. Signals collected at different time slots are recorded to formulate the CSS networks by utilizing the time diversity. Malicious users may show up in

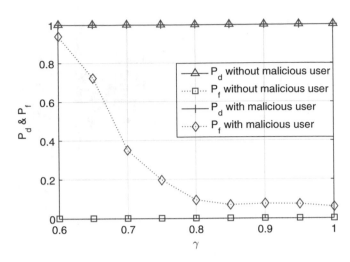

Fig. 5.11 Detection performance of the proposed malicious user detection algorithm with real signals under different compression ratio γ, rank order $K = 9$

any time slot during the signals recording period. The proposed low-rank MC-based malicious user detection framework is tested on the collected real signals.

In this case, the number of active PUs is $K = 9$. The total number of channels is $\mathscr{I} = 40$ over TVWS. The number of active SUs implemented to sense the same channel is $J = 50$. Malicious users show up in channel 27 randomly because the trial signal is set to be discontinuously over time. With the real signals, Fig. 5.11 shows the detection performance of the proposed malicious user detection algorithm versus varying compression ratios γ. The detection performance comparison is demonstrated for the cases with and without malicious users in the CSS networks. When no malicious user shows up in the CSS networks, it shows that the perfect detection performance ($P_d = 100\%$ and $P_f = 0\%$) can be achieved by choosing the suitable threshold for decision making. If malicious users show up in the CSS networks, false alarm probability becomes higher than the case without malicious users. This is because the false alarm happens if the corrupted value on TVWS channel 27 is not removed properly during the MC process. With the increasing compression ratio, the false alarm probability gets closer to the case without malicious users, as the exact MC can be guaranteed with more sensed channels at the FC.

5.5 Summary

In this chapter, a low-rank MC-based malicious user detection framework was proposed for secure CSS networks, with the purpose of alleviating the costs of data acquisition at SUs and improving the malicious user detection accuracy. Each SU only sensed a segment of the spectrum of interest. The number of active SUs in CSS networks was less than the case that MC technique is not invoked at the FC. More particularly, a low-rank MC-based malicious user detection algorithm was proposed by adopting the AOP algorithm, in which the channels corrupted by malicious users were removed during the MC process. In order to make the malicious user detection process blind, a rank order estimation algorithm and an estimation strategy on the number of malicious users were proposed to provide the required inputs for the AOP algorithm. Furthermore, the proposed malicious users detection framework was tested on both the simulated signals and the real-world signals over TVWS. Numerical analyses showed that the proposed framework achieved good detection performance with limited number of active SUs or lower costs of data acquisition at each participating SU. It can be concluded that the proposed malicious user detection framework is a good candidate for the secure CSS networks.

References

Akyildiz, I. F., Lo, B. F., & Balakrishnan, R. (2011). Cooperative spectrum sensing in cognitive radio networks: A survey. *Physical Communication, 4*, 40–62.

Boumal, N., & Absil, P.-A. (2011). RTRMC: A riemannian trust-region method for low-rank matrix completion. In *Proceedings of Advances in Neural Information Processing Systems 24 (NIPS)* (pp. 406–414)

Candes, E. (2006). Compressive sampling. In *Proceedings of the International Congress of Mathematicians, Madrid, Spain* (Vol. 3, pp. 1433–1452).

Candes, E., & Recht, B. (2009). Exact matrix completion via convex optimization. *Foundations of Computational Mathematics, 9*, 717–772.

Chen, R., Park, J.-M., & Bian, K. (2008). Robust distributed spectrum sensing in cognitive radio networks. In: *Proceedings of the IEEE International Conference on Computer Communications (INFOCOM), Phoenix, AZ* (pp. 13–18).

Ding, G., Wu, Q., Yao, Y.-D., Wang, J., & Chen, Y. (2013). Kernel-based learning for statistical signal processing in cognitive radio networks: Theoretical foundations, example applications, and future directions. *IEEE Signal Processing Magazine, 30*, 126–136.

Ghasemi, A., & Sousa, E. (2005). Collaborative spectrum sensing for opportunistic access in fading environments. In *Proceedings of the IEEE International Symposium on Dynamic Spectrum Access Networks (DYSPAN), Baltimore, MD* (pp. 131–136).

Kalamkar, S., Banerjee, A., & Roychowdhury, A. (2012). Malicious user suppression for cooperative spectrum sensing in cognitive radio networks using Dixon's outlier detection method. In *Proceedings of the National Conference Communications (NCC), Kharagpur* (pp. 1–5).

Kaligineedi, P., Khabbazian, M., & Bhargava, V. (2008). Secure cooperative sensing techniques for cognitive radio systems. In *Proceedings of the IEEE International Conference Communications (ICC), Beijing, China* (pp. 3406–3410).

Kaligineedi, P., Khabbazian, M., & Bhargava, V. K. (2010). Malicious user detection in a cognitive radio cooperative sensing system. *IEEE Transactions on Wireless Communications, 9*, 2488–2497.

Li, H. (2010). Reconstructing spectrum occupancies for wideband cognitive radio networks: A matrix completion via belief propagation. In *Proceedings of the IEEE International Conference on Communications (ICC), Cape Town, South Africa* (pp. 1–6).

Li, H., & Han, Z. (2010). Catch me if you can: An abnormality detection approach for collaborative spectrum sensing in cognitive radio networks. *IEEE Transactions on Wireless Communications, 9*, 3554–3565.

Liu, Y., Qin, Z., Elkashlan, M., Gao, Y., & Hanzo, L. (2017). Enhancing the physical layer security of non-orthogonal multiple access in large-scale networks. *IEEE Transactions on Wireless Communications, 16*, 1656–1672.

Ma, Y., Gao, Y., Cavallaro, A., Parini, C. G., Zhang, W., & Liang, Y. C. (2017). Sparsity independent sub-Nyquist rate wideband spectrum sensing on real-time TV white space. *IEEE Transactions on Vehicular Technology, 66*, 8784–8794.

Ma, Y., Gao, Y., Liang, Y. C., & Cui, S. (2016). Reliable and efficient sub-Nyquist wideband spectrum sensing in cooperative cognitive radio networks. *IEEE Journal on Selected Areas in Communications, 34*, 2750–2762.

Meng, J., Yin, W., Li, H., Hossain, E., & Han, Z. (2011). Collaborative spectrum sensing from sparse observations in cognitive radio networks. *IEEE Journal on Selected Areas in Communications, 29*, 327–337.

Qin, Z., Gao, Y., Plumbley, M. D., & Parini, C. G. (2016). Wideband spectrum sensing on real-time signals at sub-Nyquist sampling rates in single and cooperative multiple nodes. *IEEE Transactions on Signal Processing, 64*, 3106–3117.

RFeye node https://www.crfs.com/all-products/hardware/nodes/.

Tian, Z., & Giannakis, G. (2007). Compressed sensing for wideband cognitive radios. In *Proceedings of the IEEE International Conference Acoustics, Speech and Signal Processing (ICASSP), Honolulu, HI* (pp. 1357–1360).

Wang, W., Chen, L., Shin, K., & Duan, L. (2014). Secure cooperative spectrum sensing and access against intelligent malicious behaviors. In *Proceedings of the IEEE International Conference on Computer Communications (INFOCOM), Toronto, ON* (pp. 1267–1275).

Wang, W., Chen, L., Shin, K., & Duan, L. (2015). Thwarting intelligent malicious behaviors in cooperative spectrum sensing. *IEEE Transactions on Mobile Computing, 14*, 2392–2405.

Wang, W., Li, H., Sun, Y., & Han, Z. (2009). Attack-proof collaborative spectrum sensing in cognitive radio networks. In *Proceedings of the Conference on Information Sciences Systems (CISS)* (pp. 130–134).

Wang, W., Li, H., Sun, Y., & Han, Z. (2010). Securing collaborative spectrum sensing against untrustworthy secondary users in cognitive radio networks. *EURASIP Journal on Advances in Signal Processing, 2010*, 1–15.

Wang, Y., Tian, Z., & Feng, C. (2012a). Collecting detection diversity and complexity gains in cooperative spectrum sensing. *IEEE Transactions on Wireless Communications, 11*, 2876–2883.

Wang, Y., Tian, Z., & Feng, C. (2012b). Sparsity order estimation and its application in compressive spectrum sensing for cognitive radios. *IEEE Transactions on Wireless Communications, 11*, 2116–2125.

Yan, M., Yang, Y., & Osher, S. (2013). Exact low-rank matrix completion from sparsely corrupted entries via adaptive outlier pursuit. *Journal of Scientific Computing, 56*, 433–449.

Yan, Q., Li, M., Jiang, T., Lou, W., & Hou, Y. (2012). Vulnerability and protection for distributed consensus-based spectrum sensing in cognitive radio networks. In *Proceedings of the IEEE International Conference on Computer Communications (INFOCOM)* (pp. 900–908).

Zhang, L., Ding, G., Wu, Q., Zou, Y., Han, Z., & Wang, J. (2015). Byzantine attack and defense in cognitive radio networks: A survey. *IEEE Communication Surveys and Tutorials, 17*, 1342–1363.

Part III
Conclusions

Chapter 6
Conclusions and Future Work

6.1 Conclusions

This book presented research work on the promising applications of compressive sensing (CS) technique in wideband spectrum sensing, which is regarded as one of the most challenging tasks in cognitive radio networks (CRNs). It has been demonstrated that CS is capable of enabling sub-Nyquist sampling at secondary users (SUs), by exploiting the natural sparsity of spectral signals. By invoking CS technique, the signal sampling costs at SUs are significantly reduced, which is of great significance in CRNs as the SUs are normally energy-constrained devices. Within this book, the fundamental research has been presented on the design of novel compressive spectrum sensing algorithms, with particular efforts to improve energy efficiency, robustness, and security of CRNs. All the proposed designs are verified by real-world data, which also demonstrated the potential of data-driven compressive spectrum sensing.

In Chap. 3, a data-driven compressive spectrum sensing algorithm was proposed to improve the recovery performance, in which geolocation database has been invoked for providing prior information over TV white space (TVWS). By doing so, original signals were able to be recovered with requiring fewer measurements, which lowered the computational complexity of signal recovery. Simulations have been done on both the real-world and the simulated data for evaluating performance of the proposed algorithm. It is worth pointing out that SUs are capable of estimating the sparsity level efficiently by utilizing the geolocation database, which makes the proposed algorithm be more adaptive to dynamic spectrum variation. Consequently, the unnecessary energy consumption at SUs is eliminated.

In Chap. 4, through proposing new channel division schemes for single SU and multiple cooperative SUs scenarios, the amount of data sensed at SUs and data transmitted among the whole CRNs have been significantly reduced. Additionally, a denoising algorithm has been proposed to improve the robustness to channel noise. Inspired by the TVWS pilots conducted in the UK, the proposed algorithms have

© The Author(s), under exclusive license to Springer Nature Switzerland AG 2019
Y. Gao, Z. Qin, *Data-Driven Wireless Networks*, SpringerBriefs in Electrical
and Computer Engineering, https://doi.org/10.1007/978-3-030-00290-9_6

been tested on both the simulated and real-world signals over TVWS. Furthermore, numerical results demonstrated that: (1) the computational complexity of signal recovery process has been significantly reduced, and the robustness of the proposed algorithm to channel noise has been dramatically improved.

Aiming at enhancing the security in CRNs, a malicious user detection framework has been proposed in Chap. 5, by invoking low-rank matrix completion (MC) technique. The channels corrupted by malicious users are removed during the process of MC. Additionally, to ensure the malicious user detection process be independent on the prior information of networks and spectrum diversity, a rank order estimation algorithm and a malicious user number estimation strategy have been proposed. Furthermore, the proposed framework has been tested on both the simulated and real-world signals over TVWS. It has been demonstrated that the proposed framework is capable of achieving good detection performance, with limited number of active SUs or low costs of data acquisition at each individual SU.

To sum up, novel data-driven compressive spectrum sensing designs have been proposed with particular emphasis on the robustness, computational complexity, and security in this book. Amount of simulations have been done on the real-world signals over TVWS, which demonstrated that the effectiveness of applying CS in wideband spectrum sensing.

6.2 Future Work

The following three research issues have been identified and are to be addressed in the future work, for the applications and implementations of CS in wideband spectrum sensing.

1. **Performance limitations under practical constraints** Although there exist many research contributions in the field of compressive spectrum sensing, most of them have assumed some ideal operating conditions. In practice, there may exist various imperfection, such as noise uncertainty, channel uncertainty, dynamic spectrum occupancy, and transceiver hardware imperfection Sharma et al. (2016), Zhang et al. (2018a,b). For example, the centralized compressive spectrum sensing normally considers ideal reporting channels, which is not the case in practice. This imperfection may lead to significant performance degradation in practice. Another example comes from the measurement matrix design. The Gaussian distributed matrix achieves better performance but with a higher implementation cost. Even though some structured measurement matrices, such as random demodulator, with a lower cost and acceptable recovery performance degradation, have been proposed to enable the implementation of CS as a replacement of high-speed ADCs, the nonlinear recovery process limits its implementation. Therefore, it is a big challenge to further investigate compressive spectrum sensing in the presence of practical imperfection and to

develop a common framework to combat their aggregate effects in CS-enabled CRNs.

2. **Generalized platform for compressive spectrum sensing** The existing hardware implementation of sub-Nyquist sampling system follows the procedure that the theoretic algorithm is specifically designed for the current available hardware devices. However, it is very difficult or sometimes even impossible to extend the current hardware architectures to implement other existing compressive spectrum sensing algorithms. Thus, it is desired to have a generalized hardware platform, which can be easily adjusted to implement different compressive spectrum sensing algorithms with different types of measurement matrices and recovery algorithms.

References

Sharma, S. K., Lagunas, E., Chatzinotas, S., & Ottersten, B. (2016). Application of compressive sensing in cognitive radio communications: A survey. *IEEE Communication Surveys and Tutorials, 18*, 1838–1860.

Zhang, X., Ma, Y., Gao, Y., & Cui, S. (2018a). Real-time adaptively regularized compressive sensing in cognitive radio networks. *IEEE Transactions on Vehicular Technology, 67*, 1146–1157.

Zhang, X., Ma, Y., Qi, H., & Gao, Y. (2018b). Low-complexity compressive Spectrum sensing for large-scale real-time processing. *IEEE Wireless Communications Letters, 7*(4), 674–677.

Printed in the United States
By Bookmasters